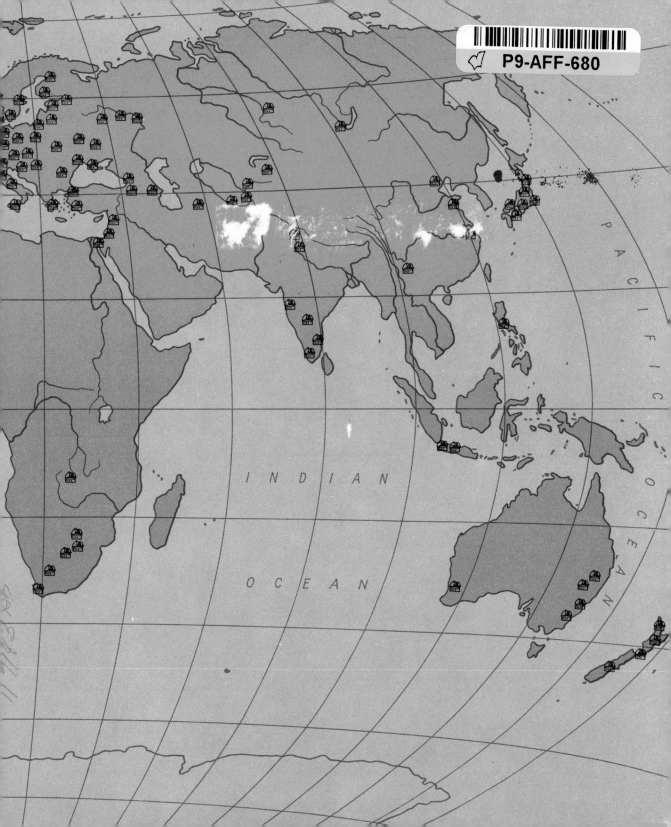

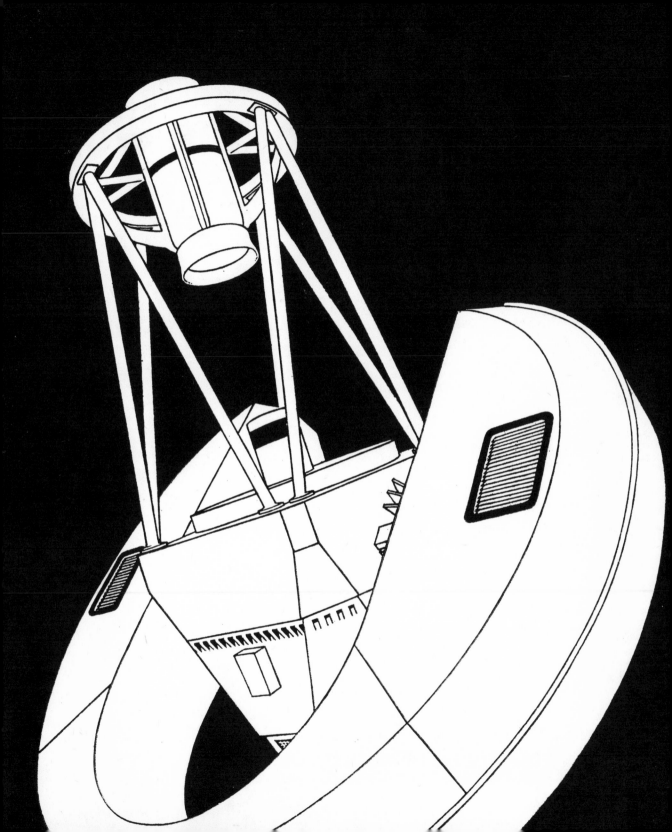

Siegfried Marx / Werner Pfau

Observatories of the World

VNR VAN NOSTRAND REINHOLD COMPANY
NEW YORK CINCINNATI TORONTO LONDON MELBOURNE

Copyright © 1982 by Blandford Books Limited

Library of Congress Catalog Card Number 81-70024

ISBN 0-442-26270-1

Originally published as *Sternwarten der Welt*
in the German Democratic Republic
World copyright © 1979 by Edition Leipzig
Translated from the German by C. S. V. Salt
Edited and revised by Dr. Simon Mitton, Cambridge

First published in the U. K. 1982 by Blandford Press
Link House, West Street, Poole, Dorset BH 15 1 LL

Published in the United States in 1982 by Van Nostrand Reinhold Company
135 West 50th Street, New York, NY 10020.

Van Nostrand Reinhold Limited
1410 Birchmount Road, Scarborough, Ontario M1P 2E7, Canada

Printed in the German Democratic Republic

16 15 14 13 12 11 10 9 8 7 6 5 4 3 2 1

Contents

The results of astronomical research have always attracted the special attention of a large number of people and they represent a major contribution to the development of the scientific description of the world. In recent years, the interest taken in astronomy has increased still further as a result of significant new discoveries such as, for example, the quasars, the pulsars, cosmic sources of X-ray radiation and organic molecules in Space, and as a result of the progress in space travel. This is reflected in the many requests addressed to observatory staffs for permission to visit the observatories and for information about the work of astronomers and the results of their research. This has encouraged us to present by means of text and pictures a number of astronomical research establishments all over the world. The scope of the book necessarily obliged us to make a narrow choice from the large number of observatories which have made major contributions to astronomical research. With the 40 institutes selected, we have endeavoured to present as many of the branches of astronomy as possible and to describe the different methods of observation and instruments employed as well as the other ancillary equipment used. We are naturally aware that the choice made is a subjective one, but are confident that we have presented a representative review of the establishments where astronomical research is carried out.

The descriptions are based on publications on the institutes in journals, on the annual reports published by many establishments and, above all, on the answers to questionnaires which we sent out to them. Without the ready cooperation of these institutes in answering the questionnaires and in providing pictorial material, it would not have been possible to produce this book.

We wish to express our thanks to the staff of the publishers, Verlag Edition Leipzig, for their collaboration and, in particular, we would like to record our debt to the late Joachim Konrad, the former Head of the Department there.

Siegfried Marx
Werner Pfau

Astronomy is one of the oldest of the natural sciences. There is evidence that it dates back to the 3rd millennium B. C. In the systematic observation of the sky, Man soon noticed that certain constellations in the sky are seen at regular intervals. This is particularly true of the brightest celestial bodies—the Sun, the Moon and the planets—which wander in relation to the stars. It thus became the practice to determine the times of religious celebrations according to the positions of the celestial bodies. It is therefore quite understandable that it was priests who practised astronomy in the early days. It was the priests, too, who in the course of time evolved methods for the reckoning of time and the calendar.

In addition to the reckoning of time, astronomical observations were soon also employed for determining geographical positions and this was especially significant for the development of long-distance trade. The determination of time and position thus became increasingly important for economic development which, in connection with the growing interest in science, encouraged astronomical observations in a systematic manner. The relation between social development and the emergence of astronomy as a science is therefore apparent. Astronomical observations have always provided the empirical foundation for our knowledge of the stars. In the past, it was their positions and movements which were the main point of interest; today, however, attention is concentrated on questions concerning the state and the development of cosmic objects.

The immense pool of knowledge, which is of particular importance for our scientific view of the Universe is the product of the close interrelation between practical observation and the theoretical interpretation of the facts retained. The subjects of astronomical research at the present time are: our solar system, that is, the Sun itself, the planets and their satellites, the asteroids, comets, and meteorites; the fixed stars as individual objects and in their conglomerations as double and multiple stellar systems; and star clusters and the stellar systems, including our Galaxy, the Milky Way system.

In addition to the more or less large bodies already listed, the objects investigated by astronomy include finely distributed matter in the form of gas or dust as well as radiations and fields. One classification scheme of this science into its various branches results from the *subject of research*. The other follows from the *methods* employed. Thus positional astronomy with the measurement of accurate stellar coordinates on the sphere provides the basis for all questions concerning the spatial distribution and movement of celestial bodies. Celestial mechanics explains these movements through the law of gravitation. In stellar statistics, the behaviour of large groups of stars is described by statistical methods. Terms such as radio or X-ray astronomy indicate the type of radiation investigated. The ultimate aim of all the efforts made in the investigation of cosmic objects is the description of the emergence and development of Space in detail and of its structure as a whole. These questions fall within the scope of cosmogony and cosmology.

There is an overlap between the classification of astronomy by content on the one hand and that by methods on the other, since positional references of very high accuracy are just as important for the examination of certain questions concerning the astronomy of the planetary system as for research concerning the fixed stars and, for example, the methods of radio astronomy are used just as much in the investigation of the Sun and Jupiter as in the examination of the physics of interstellar matter. On the other hand, the individual astronomer and in general also the research institutions concentrate on specific research subjects and methods. Contact with related fields is maintained by close scientific collaboration.

The first observatories were equipped, for the most part, with devices for angle measurement. Monuments dating as far back as the Stone Age (3000–1500 B. C.), such as Stonehenge in England, may even be regarded as angle measuring "instruments". However, these and other Stone Age structures were permanently aligned for specific positions of certain celestial bodies. For example, they enabled the exact time of the solstice to be determined or specific positions of the orbit of the Moon. For the calculation of different positions of a celestial body, movable measuring instruments were needed, such as the armillary spheres and quadrants evolved in Classical Antiquity. This development led to instruments of ever increasing size to provide scales in the form of very large circles and thus achieve a high accuracy of reading. The degree of precision achieved even in Antiquity is probably best illustrated by the discovery of the precession of the equinoxes by Hipparchus (c. 190–125 B.C.) by which stellar coordinates shift by a maximum of a bare minute of arc in the course of a year. In Antiquity, a discovery of this order was the result of a series of measurements over many years.

In astronomy, too, modern times began with a marked upswing in the 16th and 17th centuries. This is reflected in the achievements of Nicolaus Copernicus (1473–1543), who questioned the geocentric view of the world held since Antiquity by postulating the heliocentric theory; in the extensive and very precise pretelescopic observations of Tycho Brahé (1546–1601); in the work of Johannes Kepler (1571–1630) who used Brahé's observations to derive the laws of planetary motion and, especially, also in the invention of the telescope. The first telescopes for astronomical use were constructed by Galileo Galilei (1564–1642) on the basis of reports which had come to him from Holland on the subject. The development of the telescope started with the lens telescope, the refractor. There is, however, no doubt that within ten years of the invention of the lens telescope that it was already known that it was possible to form an optical image by using concave mirrors. However, the practical application of the reflecting telescope, which was pursued by Isaac Newton (1643–1727), for example, was delayed for a long time because of a lack of a suitable mirror material: this had to be highly polished and

also have a long-lasting reflection surface of a high standard. Special alloys and later glass, which was silvered chemically, then led to the reflecting telescope finding favour on an approximately equal footing to begin with. The optical image obtained with mirrors is free from chromatic aberrations (i. e. colour faults) which, in a simple lens system, are present as coloured fringes around the images and which were only really eliminated since the work of J. von Fraunhofer (1787–1826). Since with concave mirrors only the polished surface is used and the internal purity of the glass is of no importance, they can be made of greater size than lenses. Unlike a lens, a mirror can be supported at the rear so that deformations are of a much lower order. The largest refractors have a lens diameter of up to a metre and were constructed in the final decades of the last century. For technical reasons, instruments with a larger aperture, as built by F. W. Herschel (1738–1822) or the Third Earl of Rosse (1800–1867), had to be arranged as reflectors. Refractors still form part of the equipment of the world's most important observatories. However, apart from special and auxiliary telescopes, the reflecting telescope now predominates since it satisfies all reasonable requirements in respect of optical parameters, image quality and field of view diameter and, furthermore, is also employed for instruments of smaller dimensions.

It is only in combination with other optical components that a light-gathering, reflecting concave mirror becomes a telescope. The practical executions of the principle vary considerably. The choice of a certain type of optical system is made by the user on the basis of his requirements and the intended application of the telescope. Considerable expense is involved, for example, if a large field of view is specified and this is essential for an astronomer engaged in direct photography. On the other hand, there is no need for this if it is a question of the photometric or spectroscopic investigation of only a single object in each case. The astronomer engaged in the latter work requires, for

1   The Tartar prince Ulugh Beg (1394–1449) had a great observatory with very large instruments built at Samarkand in about 1420. He was not only a patron of astronomy but also carried out observations in person. He made new observations of most of the stars in Ptolemy's *Star Catalogue* and did this with a very high degree of accuracy since there were many errors in the old lists of stars.

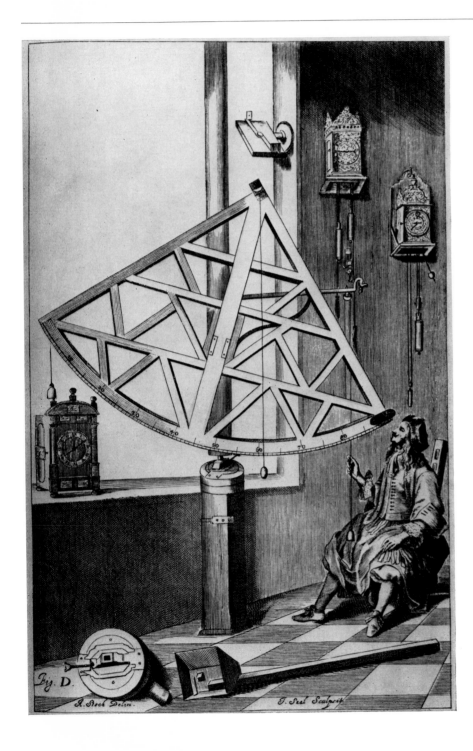

2   The positions of stars are usually measured in the horizontal coordinate system, i. e., determinations are made of the azimuth (compass direction) and of the angle of elevation above the horizon. For the observation and measurement of altitude, the azimuth mural quadrant was used in the Middle Ages and up to the beginning of modern times. The angle-measuring instrument could be moved around a vertical axis and a sight was then taken of the star to be observed along the notches on one of the arms of the right angle. The plumb-line indicated the angle of elevation on the quadrant.

3   The Multiple Mirror Telescope of the Harvard-Smithsonian Center for Astrophysics on Mount Hopkins—shown here as a model—marks the beginning of a new development in telescope design. The large light-collecting surface required is achieved by combining several individual mirrors, which, in this case, are 1.8 metre in diameter.

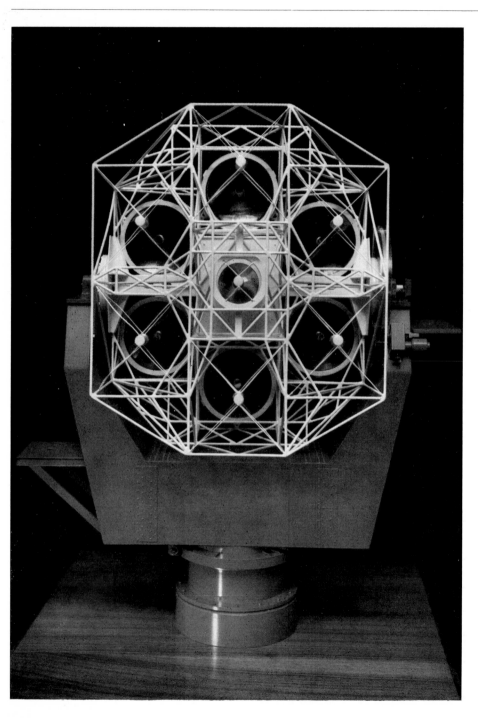

Pages 12/13:
4   This refractor is a fine example of craftsmanship and was made during the first quarter of the 19th century in the workshops of J. von Fraunhofer. He was the first to construct a parallactic (equatorial) mounting with an inclined polar axis pointing to the celestial pole which could be used for large instruments.

5   Crystalline structures form in the glass mass during the manufacturing process. As a result, there is no thermal expansion or only a very low coefficient of thermal expansion in glass ceramics of this kind. This is why they are preferred in the making of telescope mirrors. This photograph, which was taken with an electron microscope, enables the crystals to be seen.

6   The greater the focal length of the objective in relation to the focal length of the eye piece, the greater the magnifying power of the telescope. Since eyepieces with very short focal lengths were not available in the 17th century, telescopes with very great focal lengths of the objective had to be constructed in order to achieve a high magnifying power. G. Huygens, G. D. Cassini, J. Hevelius and other astronomers constructed refractors of 20 to 40 metres in length during the second half of the 17th century. An example of these is the telescope of J. Hevelius at Danzig.

7   J. Hevelius (1611–1687), an astronomer active at Danzig, as it was then known, was one of the first to use the telescope for the descriptive observation of the sky. With this engraving, which is taken from his *Machina Coelestis* published in 1673, he explains his method of observing the Sun on a projection screen.

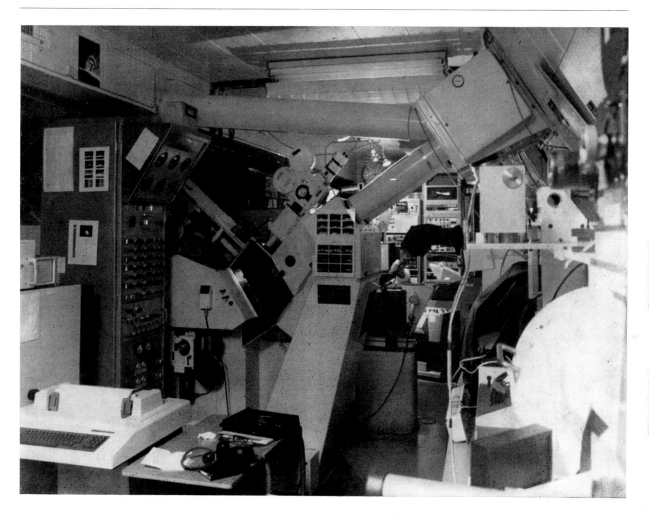

8   In comparison with Hevelius' illustration, this view of the research laboratory of Sacramento Peak Solar Observatory clearly demonstrates the effect of 300 years of technical development in optics, electronics and instrument construction.

Page 16:
9   Observations carried out from outside the terrestrial atmosphere by means of man-made satellites are of great value to astronomy since the absence of the detrimental effects of the layer of air permits observations in all wavelength regions from gamma to radio-wave radiation.

various reasons, a long focal length and more or less convenient access to his measuring equipment during the observation. The *Cassegrain telescope* is excellent for this purpose. The focus in this instrument is arranged behind the primary mirror which is pierced for this purpose and consequently offers advantages in the operation of the instruments. If the programme is concerned with high-resolution spectral observations, the spectrographs needed for this are too large, heavy, and too sensitive in respect of distortion to be swung and moved when mounted on the telescope. In this case, auxiliary mirrors are generally used to direct the light-path out of the telescope tube and through the fixed axis of rotation into a room outside the dome. In this room, it is then possible to have a considerable amount of permanently installed equipment and even to operate it in an air-conditioned atmosphere. This arrangement is known as a *coudé telescope.*

The quality of the optical image produced by the telescope is largely determined by the precision with which the final surface of the mirror agrees with the shape specified during the course of manufacture. Typical residual size errors are fractions of a light-wave length, i.e., about a ten thousandth of a millimetre. If precision of this order is to be achieved, the unworked mirror must be subjected, after casting, to a very slow cooling process in which it is reduced, free of stresses, from a casting temperature of 1500 °C and more to a normal ambient temperature. There then commences the grinding and polishing process which often lasts for months on end and which has to be interrupted, especially towards the end, by an increasing number of test procedures so that the precision attained or the corrections which still have to be carried out can be estimated objectively. The block of glass, prepared in the above manner, is finally turned into a mirror by coating it with a thin film of aluminium and, over this, by means of a vaporizing technique, a permanent layer protecting against mechanical damage. It then reflects almost 90 % of the light which strikes it.

The high level of precision in manufacture must be maintained under the conditions of night-time observation. This necessitates mirror mountings of special design in which the mirror—which weighs 42 tons in the case of the Soviet 6-metre telescope for example—is relieved of strain in such a manner that it is not mechanically distorted by its own weight. Precautions must also be taken against deformations caused by differences in temperature and the consequential thermal expansion on the front and rear faces of the mirror. The safest solution here is to use mirror materials of low thermal expansion, in which temperature differences do not have a troublesome effect on the image. This is why large mirrors are no longer made of glass but of special substances such as quartz or glass ceramics.

There are many different designs of telescope mountings. Their function is to provide a stable support and to ensure precise movement of the telescope and optical components. When observations are in progress, the optical system must accurately follow the daily rotation of the sky, the movement of the stars from east to west. Like these, it may therefore turn around an axis directed towards the celestial pole and take almost 24 hours of our normal time reckoning to complete one revolution. With this type of mounting, one speaks of a *parallactic* or *equatorial mounting*. The inclination of the polar axis and the slowness of the motion confront the telescope designer with special problems. Very slow sliding movements can lead to mechanical irregularities and the inclined axis arrangement is necessarily associated with flexing as a function of the position. It must be remembered that the tube of a 1-metre telescope weighs about three tons but has to be moved into the desired position with an accuracy of a second of arc. When the aperture is at a distance of three metres from the axis of rotation, this corresponds to a setting which is accurate to $\pm$ 15 µm.

In the past, the typical telescope mounting was the *German mounting*, as it was termed, J. von Fraunhofer being the first to use it for large instruments. In this type of mounting, the two axes—the polar axis pointing towards the pole and the declination axis perpendicular to it—can be clearly seen. Two axes of rotation are necessary for covering every point in the sky above the horizon. They are to be found in all types of telescope in one or other technical form. The *English mounting* differs from the German by having a second bearing to support the polar axis at its upper end. This provides additional stability in the case of large moving masses and sites of low geographical latitude with a correspondingly heavily inclined polar axis. In the English and the German mounting, the arrangement of the telescope tube lateral to the polar axis necessitates the use of a counterweight. This idle mass makes both kinds of mounting unsuitable for large telescopes. Symmetry, on the other hand, characterizes the *fork mounting* which is usually arranged as a simple fork. When supported at the other end in a manner similar to that of the English mounting, a frame mounting is obtained. This has the serious disadvantage that the region of the celestial pole itself directly behind the support is obstructed and cannot therefore be observed. To avoid this and yet ensure the support on both sides required under these conditions, the *horseshoe mounting* was evolved. The first full-scale use of this was for the 5-metre telescope of the Hale Observatories on Mount Palomar in California but it is now employed for almost all the great modern telescopes.

It is evident that the future course of development in telescope construction will no longer be marked by types of mounting with an inclined polar axis. This permits a simpler and less costly design and the more complicated movement of the telescope, which is a characteristic of all arrangements apart from the equatorial one, is no longer a problem in view of modern electronic control technology. A pioneering step in this direction was taken with the construction of the Soviet 6-metre telescope, the largest optical telescope in the world at the present time. For this, an altazimuth fork mounting was chosen, i.e., one of the axes of rotation

1 The stability of the mounting also exercises a major influence on the quality of a telescope. Various types have emerged, depending on the latitude of the location and on the total moving weight. In the illustration, the polar axis is depicted pointing to the upper right in each case. The declination axis, needed for the setting of a second coordinate, is perpendicular to it.

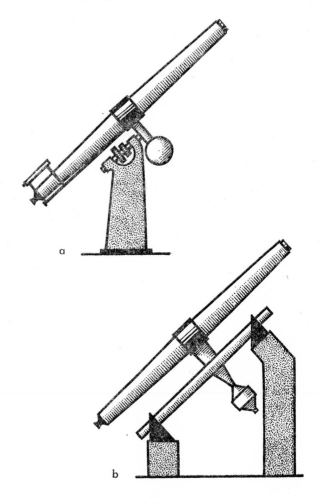

a

b

a) German or Fraunhofer
mounting
b) English mounting
c) fork mounting
d) English frame mounting
e) horseshoe mounting

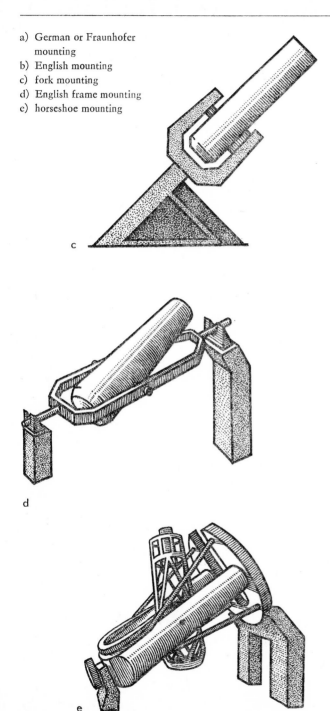

c

d

e

is perpendicular to the surface of the Earth and thus parallel to the direction of the force of gravity. To follow a star on its orbit across the sky, movements around both axes of the telescope are then necessary. The rates of movement involved are not constant but depend on the position at the particular time and the field of view turns in the course of time in relation to the sky. It is true that the control intricacies resulting from this call for the use of a computer but they do not represent a serious complication.

Very large telescopes were constructed as long ago as the 17th and 18th centuries. A refracting telescope built by J. Hevelius (1611–1687) was about 20 metres long while the 122-cm reflector of F. W. Herschel was 14 metres in length. These giant telescopes of the time were set up in the open air. However, the constantly improving optical and mechanical precision of the telescopes called for adequate protection of the instruments. Although there are various ways of enclosing telescopes, the hemispherical dome has become a typical architectural feature of observatories. The dome can be rotated and it is thus possible to turn the observation slit in any direction. In the case of a 20-metre dome, the slit is about five metres in width. Modern domes, however, are not simply to protect the precious instruments from dust and the weather but also to provide good conditions for observation in the interior of the dome. In particular, the mirrors and ancillary equipment of large telescopes can be affected by changes in temperature and this is why the radiation of the sun during the day must not be allowed to cause a rise in temperature within the dome. The easiest way of meeting this requirement is to paint the dome with a light-coloured paint which reflects a high percentage of the infrared radiation from the sun. Modern domes are also built with double walls so that the intermediate space can be well ventilated. An air-conditioning effect can likewise be achieved by a thick layer of a material having a low thermal conductivity. It may be expected that in the future the dome, so typical of observatories, will disappear, at least as far as giant telescopes are concerned, since it encloses a large space which cannot be used. A new construction technology has already been applied to the American multi-mirror telescope in which the entire building, comprising only the telescope and the rooms essential for observation, turns with the telescope.

In Antiquity, the Middle Ages and at the beginning of modern times, most astronomical observations were concerned with positional problems. They formed the basis for the determination of time and geographical position, for the elaboration of geocentric and heliocentric views of the Universe, for the derivation of Kepler's laws and for the first calculations of the distances of the fixed stars. Even F. W. Bessel (1784–1846) considered that positional measurements were the sole purpose of astronomy, although new areas of astronomical research were already emerging during his time. Astronomers needed observational instruments with which they could read off the coordinates of the stars as accurately as possible and an exact statement of time.

During the second half of the last century, a fundamental change began in the technical equipment of observatories. In 1814 J. von Fraunhofer discovered dark lines in the spectrum of the Sun. G. R. Kirchhoff (1824–1887) and R. Bunsen (1811–1899) were able to explain the origin of Fraunhofer lines and this led to the birth of spectroscopy. From a spectrum, it is possible to obtain information on velocity and physical data, such as pressure, temperature and magnetic field strength, and the chemical composition. Once again, a notable upswing had taken place in the development of astronomy—*astrophysics* had appeared on the scene.

Another discovery of significance for astronomy was also made at this time. L. J. M. Daguerre (1787–1851) succeeded in producing photographic pictures and by 1839 he had taken a picture of an astronomical object—the Moon. The photographic plate and, more recently, electronic detectors, became the most important means of recording the radiation of celestial bodies in the optical wavelength region.

Following the emergence of astrophysics and the advent of objective optical measuring techniques, the latest stage of development in this branch of science is the increase in the accessible wavelength areas obtained by radio astronomy and "short-wave astronomy". This is associated with technical developments of an entirely new kind.

The field in which an observatory specializes also largely characterizes the measuring instruments used in association with the telescopes and those instruments which are subsequently used for the evaluation of the observations. The instruments used at the telescopes are chiefly *photometers* and *spectrographs* while *comparators, plate measuring instruments* and various kinds of *plate photometers* are used for evaluation purposes.

The era when astronomers collected facts and data by the visual method, i. e., by looking through telescopes, is long since past. In our century, radiation detectors of greater reliability have taken over the task of collecting measurement data. An example of this is the photographic plate, which registers all the objects in the field of view simultaneously, retains them in a permanent manner and thus enables them to be examined again from a possibly different standpoint. In comparison with the human eye, the photographic plate is able to integrate the radiation of light sources and, as a consequence, longer exposure leads to greater range. It is therefore not surprising that nowadays practically every telescope is arranged for photographic observation. Once celestial photography, since the end of the last century, was able to retain even faint stars on photographic plates, observational astronomy experienced a tremendous upswing. The use of high-speed lenses, initially adopted from the art of portrait photography, produced a new and unsuspected picture of the sky. Not only many stars but also diffuse and extended light sources and faint irregular nebulous patches became visible. Subsequently, these have been identified as luminous interstellar matter or extragalactic star systems. To this extent, the introduction of photography

2 At the refractor (a), the light coming from a celestial object is collected by an objective and passed through the eyepiece to the eye of the observer. In the case of the reflecting telescope, the collection is performed by a concave mirror. This enables measuring equipment to be placed within the tube directly at the focal point (prime focus) or a small plane mirror can divert the light-path out of the side of the tube (Newtonian system, b). In the Cassegrain system (c), a convex secondary mirror is used to extend the focal length so that the image is formed behind the now pierced principal mirror. The coudé system (d) is characterized by a focal point which is always obtained at the same position with the aid of additional mirrors. The Schmidt camera (e) is particularly useful for celestial photography with a wide field of view. With this system, the aberrations which necessarily result with the spherical mirror employed are eliminated by a corrector plate arranged at the forward end of the tube.

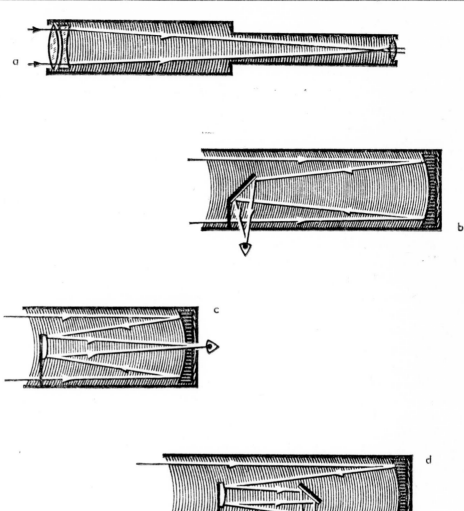

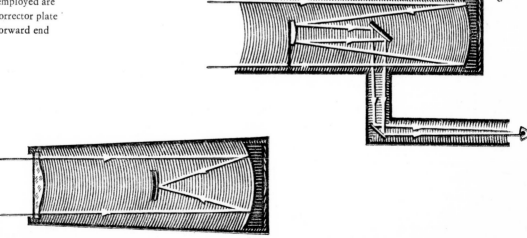

in astronomy may be compared in importance with the advance made possible by the use of the telescope to observe the sky. However, it was not only the discovery of new phenomena in the cosmos which was a notable step forward but also the accumulation of data which now became possible. This is illustrated, for example, by the statistics of the discovery of variable stars or minor planets.

At the same time, there followed a great increase in the number of newly discovered objects, due to the greater probability of discovery when examining photographic plates with images of thousands of stars in the laboratory as compared with a direct search, star by star, with the telescope. Astrometry and photometry similarly profited from the new technique of photography. The sections of the sky registered on the photographic plate are used for the derivation of exact positions of celestial bodies and for brightness measurements.

For the measurement of brightness, the photographic plate now has a serious rival in the highly developed technology of electronic radiation detectors. Sophisticated types of photoelectric cells make better use of incoming radiation and guarantee even greater range and a better accuracy of measurement. Until now, the photographic plate has still been able to counter this advantage by its capability of showing an extended image. Thus a photographic plate of the Schmidt camera at the Karl Schwarzschild Observatory at Tautenburg has a format of 24×24 cm, which can show an area of sky equivalent to sixty times the area of the full moon. 140 million separate picture elements can be distinguished on this plate and when an exposure is taken of a part of the sky containing many stars about a million stars will appear on it. The exposure time required for this, depending on the type of plate, may be less than half an hour. However, to convert the measurable densities on the plate into the brightness scale of the "magnitudes" usual in astronomy, they have to be calibrated; this is generally based on the photoelectric measurements of a few standard stars.

The two techniques are therefore complementary and it is the actual requirements of the project which determine whether the effective photographic method with its relatively less favourable internal accuracy or the exact but time-consuming photoelectric technique is used. Technical development has advanced in this sector, too, of course, and photoelectric detectors are now available which can likewise supply pictorial information over an area of sky, although the area covered is still small, and thus combine the good qualities of both types of detector. The electronic detectors are beginning to supplement or replace photographic techniques for some applications. The efforts being made in physics and in photochemistry to improve astronomical radiation detectors cannot be rated too highly. Every gain in sensitivity means greater range without having to invest large sums in bigger telescopes. It is only now, when radiation detectors are approaching the physical limits of their capabilities that further advances in the detection of even weaker light sources are dependent on new observation techniques or telescope concepts.

Many people incorrectly consider that the work of the astronomer is done only during the hours of darkness. Admittedly, the fundamental observations are usually made at night but these only become well-established knowledge when they are duly prepared or have passed through a process of theoretical assessment. These represent important aspects of an astronomer's work. There are locations which are characterized by favourable weather where an astronomer really can concentrate, night after night, on the gathering of data for quite a time. However, this is always followed by an evaluation phase which lasts for a fair time, usually many months and perhaps years.

The direct result of a night spent in observation is not at all suitable for immediate further use at the desk. The material may consist variously of photographic plates which show the celestial bodies in the form of a picture, or of a spectrum, or of recording curves, or a collection of numerical data, as is the case with photoelectric photometry or spectrophotometry and in the measurement of coordinates at the telescope. It is from these sources that the information required for the specific task must be extracted and processed in such a manner that all the influences of the particular observation technique and the instrument employed are eliminated—and this also includes the effects of the terrestrial atmosphere. A set of data is then obtained which is "absolute" in the sense that it can be communicated to other astronomers without having to refer to the actual measuring procedure. For this reason, the inventory of an astronomical observatory includes not only the telescopes and their measuring equipment but also a selection of laboratory equipment. The same careful attention is paid to the technical development of the latter as to the instruments in the dome. It is a question of obtaining data without additional errors of measurement and, through automation, of achieving high efficiency in the use of telescope time. The latter point has a direct influence on the intelligent use of telescopes since a valuable clear night is only used to the best effect when the results of the observations are rapidly made available for scientific applications. Thus automation and computer technology, as in all branches of science, also play a major role in astronomy. Their influence begins with the acquisition of observation material at the telescope, but they also speed up the process of evaluation and, finally, they are a great help in the theoretical processing work. The individual applications include:

– the positioning and tracking with the telescope, the monitoring of the dome,
– the direct control of measuring equipment attached to the telescope, the acceptance, storage, and initial data reduction on-line operation,
– the on-line operation of laboratory equipment,
– the preparation of measurement data by reduction and correction programmes,
– working out mathematical expressions during the theoretical treatment of the material,
– the use of computer memories as a data base and literature catalogue,
– assistance with the layout of the printed manuscript for the publication of scientific work.

These wide-ranging tasks are undertaken with the aid of: *microprocessors* which are built into individual items of measuring equipment and have a fixed programme for this purpose; *process computers* which carry out numerous functions and can control several items of equipment at the same time with account being taken of certain priorities; and *data-processing units* or scientific computers with a high calculating speed which are either centrally located in the institutions or are linked to these by long-distance data lines. It is precisely on account of the three latter points that computer technology is of great importance for researches concerned exclusively with theoretical questions since it is now possible to investigate problems which in the past were quite out of reach.

A characteristic feature of astronomical research in the second half of this century is the expansion of the spectral range which is accessible to observation and thus to well-founded theoretical interpretation. In the investigation of light in respect of direction of incidence, intensity and, finally, spectral composition, astronomy had to rely initially on the human eye as a "radiation detector". With the advent of radio astronomy, a "window" was opened through which vital, previously unsuspected, and optically inaccessible information on the state, distribution and movement of cosmic matter was obtained. Radio astronomy deserves a great deal of the credit for the present "explosion of knowledge" in astronomy. This is the result of a breathtaking technical development which, within a few decades of the chance discovery of cosmic long-wave radiation by K. G. Jansky in 1932, led to the post-war use of the radar installations of the Second World War and to the great radio observatories with their mighty aerials and their advanced electronic systems.

Since cosmic bodies not only emit radiation in the form of light or in the range of our radio waves but also cover the entire spectrum from gamma radiation and X-rays, through ultraviolet and visible radiation, down to the infrared part of the spectrum and the long waves, there are specialists working on all these types of radiation. Apart from the visible region, the adjacent infrared portion and parts of the radio spectrum, however, observations can only be made from locations at a great height above the surface of the Earth or from outside the terrestrial atmosphere altogether, since the layer of air around us acts as a protective screen blocking out other radiation components. This has led to an extension in the technology of observation and to the application of the term "observatory" for high-altitude balloons and astronomical research satellites. Further spectacular developments are to be expected from the latter in particular. These observation platforms are also of great value to astronomers operating in the optical range. This is because the image obtained is not disturbed by the blurring effects of atmospheric turbulence; therefore very many more details can be identified on extended bodies. In the investigation of star-like objects, the absence of atmospheric turbulence permits a concentration of light at least ten times greater in the focal plane of the telescope, thus enabling light sources of very much less brightness to be examined. A realistic projection states that it will be possible to cover up to ten times the distance which is within range at the moment with Earth-bound technology. Nobody can tell at the present time, how much this will widen our horizon.

From the above remarks, it is clear that the state of astronomical research is always dependent to a major degree on the level of technological development. While this is of general validity, the economic performance of a country also plays an important part in its level of astronomical research since the construction and development of observatories is associated with constantly increasing investments.

The construction and development of observatories in the modern sense began in the 18th century. A detailed investigation (D. B. Herrmann) has been made of the observatories founded since this time: an exponential increase in the number of observatories took place between 1800 and 1900. They doubled in number between 1800 and 1850, again by 1880 and once more by 1900. This was associated with the economic development of those years which stimulated and thus promoted progress in science.

The establishment of observatories in the various countries did not at all follow the same path. A greater number of observatories were built in those countries which held a leading economic position. One consequence of this was that up to the mid-20th century almost all observatories were to be found in the Northern Hemisphere. The construction of great astronomical research centres in the Southern Hemisphere, often as joint projects by several different countries, began after 1950 on a large scale. In the choice of location, great importance is usually attached today to outstanding local atmospheric conditions. It must also be guaranteed that the working conditions for the astronomers do not deteriorate through adverse industrial developments in the vicinity.

For the future, too, the aim of observational astronomy will continue to be the detection of low-luminosity bodies and the improvement of the resolving power of telescopes. This is equivalent to covering greater distances and thus penetrating further and further the past of the Universe. This objective calls for telescopes with larger and larger apertures.

The largest telescope in existence at the present time is the 6-metre reflector in the Soviet Union. If this size is to be exceeded to any great extent, an entirely new approach to telescope design must be found. It has been proposed that mirrors could be constructed from several elements in a mosaic fashion. In the USA there is a definite project to build a 10-metre telescope in this manner. There are even bold plans for a future 25-

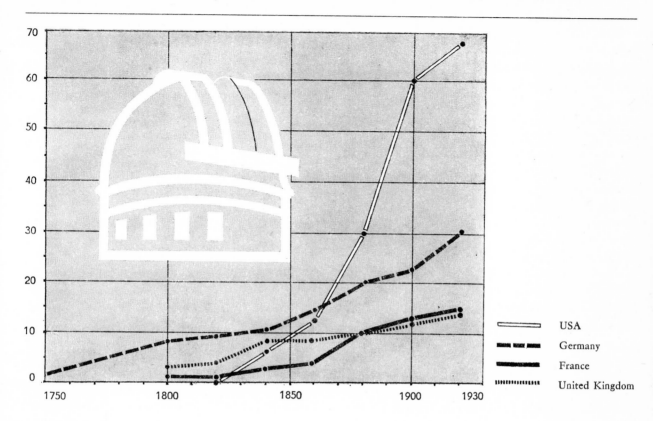

70

60

50

40

30

20

10

0

1750   1800   1850   1900   1930

	USA
	Germany
	France
	United Kingdom

3 The number of observatories in the various countries as a function of the available scientific potential demonstrates notable differences over a period of time. This is shown here from the examples of Germany, United Kingdom, France and the USA in the period from 1750 to 1930. (After D. B. Herrmann)

metre telescope which would then be equivalent to a light-collecting surface of seventeen 6-metre mirrors. The chief problem with mosaic mirrors is that very precise adjustment of the individual elements must guarantee an ideal reflecting surface in any position of the instrument.

Another possibility is to collect the light of several separate mirrors at a common focus. In this approach, the individual mirrors can be located in a common mounting, as has already been done in the American multi-mirror telescope. In this case, too, the problem is the precise gathering of the different light-paths at a common focus. For a 25-metre mirror of composite elements, 150 2-metre mirrors or seventeen 6-metre mirrors would be needed. This would mean that the mounting would have to meet very high stability requirements. It has therefore been envisaged that several

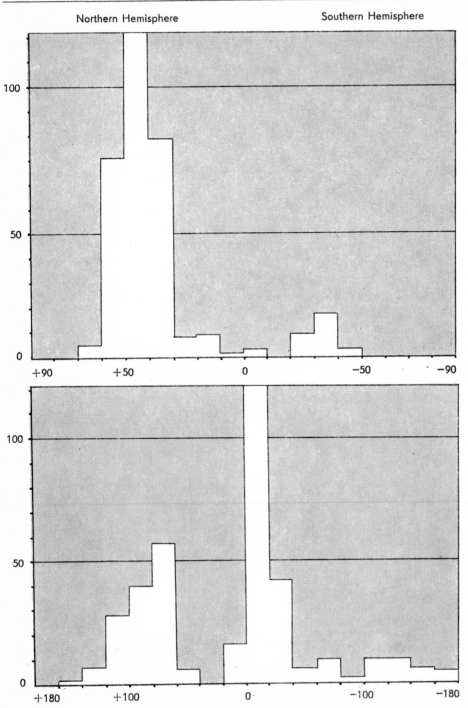

4 a/b The distribution of observatories, as given in *The American Ephemeris and Nautical Almanac* for 1977, according to geographical latitude, clearly demonstrates that they are far more numerous in the Northern Hemisphere than in the Southern Hemisphere. The picture takes on a different aspect when only the observatories with giant telescopes having a reflector diameter of more than two metres are considered. The difference is then no longer so great since in this case only modern installations dating from the immediate past are present.
The distribution of observatories according to longitude shows two centres of greater density, one in Europe and the other in North America.

5 Mirrors for astronomical telescopes have a maximum potential diameter of only about 8 metres when made in a single piece. To obtain larger light-collecting surfaces, the light-paths of several individual telescopes must be concentrated at a single point. A battery of telescopes of this kind is known as an array. To obtain the equivalent of a 25-metre telescope, 150 2-metre mirrors, 40 4-metre mirrors or 17 6-metre mirrors are required.

"classical" telescopes could be set up next to each other and controlled by a common computer centre with the light-paths being concentrated on a single point, such as for example, a spectrograph. This type of arrangement is referred to as an *array*. The economic advantage in comparison with the multi-mirror telescope consists in the possible setting up of a production line for the individual telescopes. However, there is an additional advantage in the parallel operation of separate telescopes which has long been exploited by radio astronomy. Two separate reception arrangements can be regarded as two points of a telescope having a diameter equivalent to the distance between these two individual telescopes. The diameter and thus the resolving power would thus be considerably increased without, however, involving an increase in the light-collecting surface, i. e., the aperture.

These ideas on the further development of optical astronomy are worthwhile since the majority of the observations in this wavelength area will always be made from the Earth. The extraterrestial possibilities will primarily be utilized for the detection of radiation components which are beyond the range of Earth-bound observers.

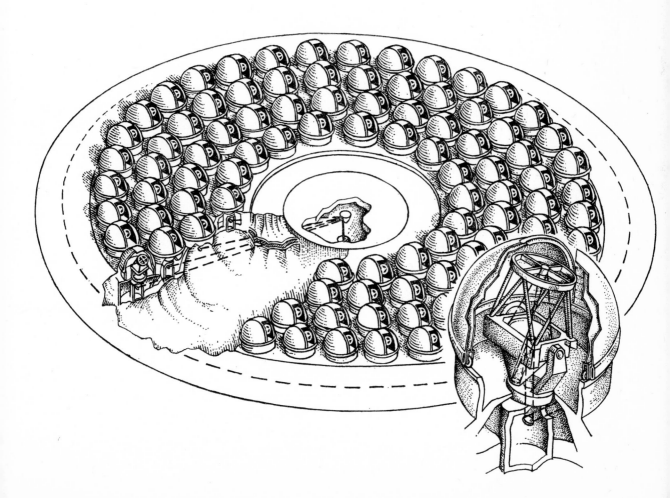

● Right in the South of the Soviet Union and not far from the frontier of the Georgian Soviet Socialist Republic with Turkey, the Abastumani Observatory of the Georgian Academy of Sciences is situated in a beautiful part of the landscape. The buildings are scattered over the wooded slopes of Kanobili, a mountain rising to 1,700 metres. The Observatory is linked by cable railway with the town of Abastumani, long famous as a health resort.

It was in this region, towards the end of the last century, that S. P. Glasenap, a Russian astronomer, observed double stars with the aid of a small refractor. The comprehensive nature of his results and their accuracy led the well-known American researcher of double stars, S. W. Burnham, to remark that ". . . no Observatory in Europe has so favourable a location . . ." and he also expressed the expectation that ". . . doubtless the Russian government will place him (i. e., Glasenap) in a position to carry on with more powerful instruments the work inaugurated at Abastumani." Despite private collections, however, the plans to build a larger observatory in this region came to nothing. It was only after the overthrow of the Tsarist Government that this plan again became feasible. The actual work on the construction of the first mountain observatory of the Soviet Union with the first telescope to be built in that country, a reflector with a mirror diameter of 33 cm, began in 1932. It was a far-sighted decision not to place the telescope in Glasenap's old observation tower in the immediate neighbourhood of Abastumani but to build the observatory where there were even more favourable conditions and, in particular, at a good distance from civilization, on Mount Kanobili.

The difficult years of the Second World War did not prevent further work from being carried out on the observatory and astronomical research from being done there. Under these difficult conditions, the staff developed the ability of achieving noteworthy results with simple resources. Thus, in the course of time and with the aid of a 40-cm refractor, a relatively modest instru-

ment, there appeared comprehensive catalogues with photometric measurements of magnitudes and colour indices, spectral types and luminosity classes of thousands of stars, which provided information about galactic structure and interstellar matter. The acquisition of such fundamental data for astrophysical research was not done haphazardly but according to the Parenago Programme, named after its initiator. Numerous investigations concerned interstellar matter: not only with matter in its general and diffuse distribution but also in the special forms such as dark clouds, gas nebulae and planetary nebulae, where gas and dust appear in a more concentrated form. Constant improvements were made in the methods employed for this work. Initially, exclusive use was made of photographic techniques. These were supplemented by photoelectric measurements of starlight and of its state of polarization. Of course, other telescopes were used in addition to the 40-cm refractor. The measuring equipment required was constructed in the Observatory's own electronic laboratory. Other important projects are concerned with the definition of a reliable classification of stellar spectra, the observation of variable stars and statistical studies of the distribution of stars in the Milky Way system.

Of the more recent telescopes on Mount Kanobili, two deserve special mention. One of these was built in 1955 and is designed for the photographic observation of stars over a wide field. It is thus equivalent to the Schmidt camera but in this case the optical principle was evolved by the Soviet optics expert D. D. Maksutov. In comparison with the Schmidt system, it is characterized by its greater ease of manufacture and shorter overall length, these being two important economic aspects. The telescope is equipped with various objective prisms which permit the simultaneous exposure of a large number of stellar spectra on a single photographic plate. The biggest telescope for astrophysical research at Abastumani Observatory at the moment is a 125-cm reflector which has only recently been installed. It is used principally for photometric and spectroscopic work; the use of television techniques is also being examined for certain investigations. There is provision for fully automatic operation of the instrument with the aid of a computer.

Scientific collaboration, which has always been important in astronomy, is carried on by Abastumani Observatory not only with partners inside the Soviet Union but also with the observatory of Helsinki University. A very powerful photometer was developed and built there, this being capable of receiving photometric data in five wavelength areas simultaneously. This instrument is permanently linked to the 125-cm telescope while in return the Finns have several months' observation time per year at their disposal in these favourable climatic conditions.

The staff of Abastumani Observatory, numbering some 40 persons, are not only concerned with galactic objects since intensive solar research is also carried out. Without exaggeration, it can be said that from Mount Kanobili the sky is kept under observation day and night. The monitoring of the Sun for the occurrence of solar activity dates back to the time when observations were first started in the 1930's. Nowadays, research methods and instruments are far more sophisticated so that the different layers of the solar atmosphere can be investigated in a systematic manner. Optical methods are supplemented by radiation measurements in the area of radio waves.

The research pursued on Mount Kanobili also includes the Moon, the planets and the higher atmospheric layers of the Earth. The aim of this work was and is to obtain a deeper understanding of the dependence of the physical chemistry of the upper layers on solar ultraviolet and particle radiation. One way to achieve this is by the observation of the emission spectra of the atmospheric gases occurring in night sky light.

The scientists of Abastumani keep in close touch with the astrophysicists at the university of their capital, Tbilissi. A technical laboratory of the Observatory has

been established there and some theorists are based in that city. In particular, however, staff from both institutions take part in training would-be astronomers and in disseminating a knowledge of astronomy among the general public.

● The history of Asiago Observatory began at Padua in 1768. G. Toaldo founded an astronomical observatory at the University of Padua and was its first director until his death in 1797. The Observatory was accommodated in buildings erected in 1768 around an old tower dating from 1242. It was only in 1942 that this university institute led to the establishment of Asiago Observatory close to the town of the same name. Nevertheless, the institutes in Padua and Asiago still constitute one scientific unit. The project in Asiago was initiated by G. Silva, who succeeded in obtaining for the Observatory, right from the start, a reflector of 1.22 metres, a large instrument at that time. A Schmidt camera with an aperture of 67 cm was acquired in 1967. This instrument was needed for a new field of activity of the research centre which is still characteristic of Asiago—the search for supernovae and the investigation of the nova and supernova phenomenon.

Every observatory not recently established—and this is the case with Asiago, of course—is confronted with the same problem. Asiago, a small town with 5,000 inhabitants, grew in extent because of its popularity as a tourist centre and the many new hotels built as a consequence of this. The result is that Asiago Observatory, established only in 1942, acquired a new outstation in 1973, situated on Mount Ekar, 1,366 metres high, which however is only four kilometres to the east of Asiago. A 1.82-metre reflector was brought into service here in 1973. 1973 marked the 500th anniversary of the birth of Copernicus and since the originator of the heliocentric view of the Universe had once studied at the University of Padua, the new instrument was given his name. He was thus given a fine memorial at the place where he had been a student. Several manufacturers were associated with the construction of the Copernicus Telescope. The mirror is of Schott Duran 50 glass which has a low coefficient of thermal expansion. The mechanical system was supplied by an Italian firm which had already been involved in the building of the Schmidt camera. A large-capacity computer is available for the electronic control of the 1.82-metre telescope. Within the next few years, the Institute intends to relocate all its observational instruments on Mount Ekar since there are better atmospheric conditions there for successful astronomical observations.

From 1768 to 1942, the research work carried out at Padua Observatory mainly concentrated on the problems of classical and geodetic astronomy. It was only with the construction of the 1.22-metre telescope and the associated move to Asiago that this changed. Astrophysical problems, such as the investigation of the variable stars, particularly novae and supernovae, the study of galactic nebulae and galaxies then began to dominate the scientific work there. This is also apparent from the ancillary equipment of the telescopes.

Powerful instruments with large fields of view are needed to find and investigate novae and especially supernovae. Up to now, only three supernovae have been optically discovered in our Milky Way system, all the other being in extragalactic stellar systems. The three galactic supernovae are the one observed in 1054 by Japanese and Chinese astronomers, Tycho's star of 1573 and Kepler's supernova of 1604. It is possible that a new star observed in the Orient in 1006 was a supernova, but this is not firmly established. The great distance of all other supernovae causes them to be observed as faint bodies although when a supernova occurs a star can increase in brightness by 20 magnitudes which is equivalent to an increase in luminosity of about a million times. However, when this happens in a galaxy which is a hundred million light-years away, the supernova at maximum brightness is only a body of about the 15th or even 18th magnitude. Powerful instruments like Schmidt cameras, as installed at Asiago with apertures of 40 cm and 67 cm, are used for the observation of distant galaxies and the supernovae which occur there.

However, it is not just a question of following the development in brightness of the supernovae that interests astronomers but also, as far as possible, of investigating the spectral characteristics. This is particularly difficult with faint bodies of that kind. At Asiago the Schmidt cameras are equipped with objective prisms for this purpose. With the objective prism of the "Small Schmidt" instrument, a dispersion of 45 nm per millimetre at the H-gamma line of hydrogen is achieved, this figure being 65 nm per millimetre at the H-gamma line with the prism on the 67-cm telescope. A distinction is drawn by astrophysicists between two types of supernova according to the light-curve and the spectrum. With a type I the continuous spectrum is very faint in the ultraviolet spectral region and with a type II it is very strong in this region. Asiago Observatory is making a great contribution to the investigation of this interesting phenomenon and the new 1.82-metre telescope with its grating spectrograph, its image intensifiers and its photometer, is being used especially for the spectrographic and photometric investigation of peculiar galaxies, quasars, novae, supernovae and galactic nebulae (see fig.6).

● Archenhold Observatory at Berlin-Treptow is known for its extensive activities in the field of popular science, for its research work on the history of astronomy, and for its outdoor telescope of 21 metres in length. There is a close link between the construction of the great telescope and the founding of this observatory in Berlin.

On 27 October 1891, F. S. Archenhold working at the photographic outstation of Berlin Observatory succeeded in taking pictures of nebulae, including the nebula at ξ Per in the constellation Perseus, with portrait lenses of short focal length. Although the nebula was already known at the time, this was the first picture of its actual extent. As a consequence of this work, F. S. Archenhold planned the construction at Berlin of a great photographic telescope exclusively for the investigation of cosmic nebulae. He envisaged an objective with an aperture of 125 cm and a focal length of 5 metres. However, it quickly became apparent during preliminary discussions with E. Abbe and F. O. Schott that the project was not feasible. The instrument which was finally built had an objective of 70 cm diameter and a focal length of 21 metres. The mounting of such a giant telescope was an enormous technical problem, especially the requirement that the pivot of the telescope and the view into the telescope, i. e., the focus, should coincide and that the instrument should be set up in the open air without a costly dome but nevertheless be protected

from the effects of the weather. A young designer from a Berlin engineering factory developed a fork mounting weighing 130 tons which was provided with numerous compensating weights, this arrangement being known as a Meyer mounting after the man who had conceived it. It thus proved possible to site the eyepiece of the instrument at the intersection of the declination and polar axes, i. e., at the pivot point of the great telescope.

The Archenhold Telescope was set up in Treptow Park in Berlin on the occasion of the Berlin Industrial Exhibition inaugurated on 1 May 1896. The first observations with this giant telescope were carried out in September of the same year.

F. S. Archenhold had intended that the telescope should be used for purely scientific tasks and that when the Exhibition was over, it should be moved to a site which was more favourable for astronomical observations. However, even when it was first set up, he already had in mind the establishment of an astronomical museum to illustrate the historical development of observatories and telescopes. A wooden building which was erected for this purpose also contained lecture rooms for the dissemination of astronomical knowledge. In October 1896, the decision was taken to keep the telescope in Treptow Park after all and to use it, too, for the purposes of popular science. This was the beginning of the People's Observatory at Treptow.

It was not long before the space available proved insufficient for the large numbers of visitors. This is why the old observatory was demolished in the spring of 1908 and the foundation-stone of a new building laid on 27 May of the same year; the official opening of the new People's Observatory at Berlin-Treptow then took place on 4 April 1909.

F. S. Archenhold was the director of Treptow Observatory from the time it was opened until 1931 and it was named the Archenhold Observatory in his honour on the occasion of the 50th anniversary. After the Second World War, numerous new instruments were installed in the Archenhold Observatory, including

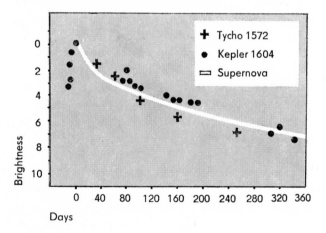

6 The diagrammatic lightcurves clearly demonstrate that supernovae can be distinguished according to the course of their brightness in the brightness fall-off phase. The decrease in brightness is appreciably more abrupt in the supernovae of type II. The brightness measurements for the supernovae of Tycho and Kepler plotted in these diagrammatic curves show that both cases were supernovae of type I.

a 50-cm Cassegrain reflector, a 15-cm coudé refractor, a 25-cm comet seeker and a 12-cm astrograph.

Even during Archenhold's time education was the principal activity of the Observatory and the dissemination of astronomical knowledge at the popular science level is still one of its major tasks. It was in connection with this that F. S. Archenhold suggested as early as 1924 that a planetarium should be built in the immediate vicinity of the Observatory. He was not able to implement this project but it finally became reality in May 1959 when a small planetarium was opened.

One of the typical facilities for the explanation of astronomical ideas is the Solar Physics Room which was established in 1966. An image of the Sun, measuring between 80 cm and 3 metres, is projected into a lecture room by a heliostat and it is also possible for a line spectrum of the Sun to be reflected into the lecture room with the aid of a four-prism spectrograph. The spectrum is 40 cm in height and its overall length is three metres. The figure of 40,000 visitors annually is evidence of the popularity of the Archenhold Observatory. The extensive programmes of the Observatory include guided tours, lectures, night observations, exhibitions and regular study-groups. In addition to the improvements carried out to the equipment of the Observatory, a great deal of exhibition and educational material has been collected and compiled; this is used for the popularization of astronomical research results. The Observatory has a collection of historical astronomical instruments such as telescopes, sun-dials, and celestial globes. In the library, there are works of great historical value and also bibliographical rarities. The comprehensive collection of books constitutes a most important basis of the scientific work at the Archenhold Observatory which is totally dedicated to the history of astronomy.

● Since the middle of the last century, the name of the city of Bonn has been associated with the *Bonner Durchmusterung* (Bonn Survey), a term familiar to every astronomer. This survey was published by F. W. Argelander (1799–1875) and lists the 324,000 stars and their locations as identified by him and his two co-workers with the aid of a very small telescope. All these stars are marked on maps, and, together with their coordinates and apparent magnitudes, are listed according to a special system of classification in catalogues. They are still used by astronomers today as a means of identification for certain stars. The more than 700,000 separate measurements made in the course of seven years were obtained by the observer at the telescope calling out the moment that a star passed across a mark in the field of view to an assistant so that the latter could note the transit time. This was subsequently converted into one of the two coordinates of the star, the right ascension. The other coordinate, the declination, was noted by the observer himself, together with an estimate of the brightness of the star. Although up to 30 stars per minute had to be recorded in areas with a high star-density, the number of incorrect measurements was exceptionally low and the entire work is a model of good planning, diligence and reliability.

F. W. Argelander was initially the director of the University Observatory at Turku and then at Helsinki, after the University had been moved there. He subsequently took up an appointment at Bonn University and founded an observatory there between 1839 and 1845. It was here that he made two contributions to astronomy which are still actively associated with his name: a method for the visual assessment of the brightness of variable stars and the *Bonner Durchmusterung* referred to above.

After Argelander's old observatory had become totally enclosed by the city of Bonn and since there was no longer enough room available there, the astronomers moved to a modern new building in the early 1970's. They are accommodated there, under one roof, with the other astronomical institutions of the University, i. e., the Radio Astronomy Institute, the Institute for Astrophysics and Extraterrestrial Research, and the Radio Astronomy Special Research Department. The Max Planck Institute for Radio Astronomy also occupies the same building.

A new optical observing station of the University Observatory was established in the Eifel district as long ago as 1954 and is known as the "Hoher List" outstation. The environmental conditions there are good since the Maare area of the Eifel plateau is part of a national conservancy scheme and consequently there is no risk of disturbances from new industrial or residential housing projects. The little town of Daun is 5 km away from the Observatory and is also to the north of it, i. e., not even in the main observation direction. The principal instrument at "Hoher List" is a Cassegrain telescope with a free aperture of 106 cm and a focal length of 15 metres and it is used for photometric and spectroscopic work. Together with photographic positional astronomy, these are the two main fields of activity as regards practical observation at the University Observatory in Bonn. Both the photometric and spectroscopic investigations are concerned primarily with the interesting subject of variable stars, a matter of great importance for the theoretical understanding of stars. Modern electronic techniques, such as the image-intensifying technique, are employed to obtain the spectra of stars even of very low brightness. The variations in the light emitted by stars was a subject which was investigated by F. W. Argelander and positional astronomy is another subject which has a long tradition at Bonn. It is being continued at the present time in the derivation of the proper motions of stars in star clusters from photographic exposures taken at two widely separated epochs with the aim of deriving spatial movements.

⊕ To increase the working facilities at Bonn Observatory, the first plans for carrying out radio observations were drawn up in 1951. Radio wavelengths are not affected by cloud or weather conditions and research in this sphere began all over the world at this time. The efforts made led in 1956 to the commissioning of a radio telescope of 25 metres in diameter—a noteworthy size for the time—on the Stockert near Bonn. The experience gained with this instrument provided the basis for the establishment of a great radio astronomy institute, the Max Planck Institute for Radio Astronomy as it is now known. The seat of the Institute is in a suburb of the city of Bonn and the main instrument is a telescope of 100-metre reflector diameter which was taken into preliminary service in the summer of 1972, the period of construction having lasted a good four years. This fully steerable aerial installation, the largest of its kind in the world, is at Effelsberg, 40 km south-west of Bonn. This location is screened to a certain extent from terrestrial radio traffic and, being in the hollow of a valley, is protected from high winds.

The magnitude of the technical achievement represented by this telescope can be appreciated if one imagines a steel structure the size of a football field which is fully steerable and can be brought from a horizontal position to an almost vertical one. Fluctuations in the adjustment and follow-up accuracy when tracking cosmic radio sources over a long period do not exceed six seconds of arc, corresponding to a deviation at the edge of the "football field" of only 1.5 mm. Another important requirement which had to be met was that the waves had to be concentrated at the focus which meant that the surface of the reflector had to be the shape of a paraboloid whatever position it was in. This applies not only to the large-area shape since even over small sections the deviations from the mathematical ideal of the paraboloid must not exceed certain limits. If this were the case, the incoming radio waves would be scattered by the "irregularities" and radiation losses

7 a/b The illustration shows two corresponding sections from the catalogue and chart of the *Bonner Durchmusterung* which was produced between 1852 and 1859 under the direction of F. W. Argelander. To convey an idea of size, a scale image of the crescent of the Moon is included. In the catalogue, the stars are classified according to their coordinates, arranged in groups of ten and numbered consecutively in zones. A clear identification of the objects is thus possible. For orientational purposes, the first column of the catalogue indicates the approximate brightness of the stars. At the

+40°   **2ᵘ—3ᵘ**

501—560					561—620					621—680					681—740					741—800				
m	2ᵘ		+40°		m	2ᵘ		+40°		m	2ᵘ—3ᵘ		+40°		m	3ᵘ		+40°		m	3ᵘ		+40°	
7.6	13	53.5	48.9	K	9.0	28	40.9	7.9		9.5	46	38.3	45.1		8.9	0	43.4	14.3		9.2	16	12.0	47.8	B
9.5	14	19.0	12.5		9.0		50.3	24.7	B	8.7		42.4	34.8	K	9.3	1	31.0	54.5		9.5		13.4	6.1	
9.3	15	1.2	57.1		8.9	29	51.5	42.9		9.5		51.0	24.3		8.8		57.3	47.5		9.2		36.3	13.5	
9.3		10.0	26.1		9.3		56.4	13.7	R	8.6	47	0.6	52.1	K	7.8		59.2	37.2	K	8.8		39.8	29.0	
9.4		10.5	13.4		9.5	30	8.0	42.0		8.9		8.9	52.4	K	9.5	2	5.0	5.8		9.4		54.3	3.4	
9.5		17.1	17.4		9.1	31	3.5	56.6	B	9.5		52.0	12.7		8.5		29.9	53.0	K	9.1	17	4.9	18.6	
9.3		18.9	7.2		8.7		23.5	42.3	K	9.3		57.4	22.5		8.7		52.8	56.7	K	8.8		10.6	17.7	B
9.5		57.0	29.1		8.3	32	20.8	16.5		8.7	48	21.3	23.1	K	9.5	3	2.6	8.0		8.3		12.9	5.3	K
9.5	16	8.0	49.2		9.5		22.6	30.5		9.3		23.0	35.5		9.3		2.9	5.8	B	9.4		23.0	29.3	B
9.1		29.0	42.0	K	7.2		48.0	47.2	K	9.3		53.7	43.6		9.4		10.2	4.8	.	8.5		23.2	45.8	K
9.5		38.7	27.5		9.5	33	1.2	6.2		9.3		55.7	30.3	K	8.2		13.4	53.5	K	9.3		42.5	33.0	B
9.5		42.5	25.4		9.5		8.3	56.4		9.5	49	32.4	30.4	B	9.0	4	9.7	38.5		9.5	18	11.2	14.6	
8.3		58.0	14.5		8.6		9.0	37.8		9.5		39.6	30.4		9.5		14.3	14.4		9.4		15.3	6.4	
8.2	17	8.5	13.5	L	9.5	34	4.7	7.2		9.2		48.5	4.6		8.8		32.7	20.8		9.5		24.6	27.1	
9.2		9.3	54.9		9.0		9.0	1.5	K	9.1		54.9	50.1		8.2		33.8	54.3	K	9.5		27.8	51.6	
9.1	18	3.0	40.0		8.5		14.0	29.0		9.5		58.9	27.4		9.5		40.6	20.8		9.5		32.5	56.0	
8.2		15.6	52.0	B	7.1		29.8	52.7	B	9.3	50	6.7	35.4		9.4	5	28.1	54.6		9.5		33.2	46.1	
9.4		18.5	10.8	B	9.3		38.2	29.0		9.4		14.7	55.8	B	9.0		41.7	37.1	K	9.5		53.8	31.5	
9.5		19.4	5.3		9.1		55.8	17.6		6.5		19.0	26.7	K	9.0	6	29.6	27.0		9.5	19	4.3	12.8	
7.9		22.5	17.1	B	9.5		57.5	23.5		9.0		23.8	6.6		9.5		53.2	59.5		9.4		11.8	58.6	
7.7		27.0	23.5	L	9.0	35	2.2	8.9		8.4		27.4	14.7	K	9.5	7	22.8	24.8		9.5		14.9	16.5	
8.8		42.5	49.6		9.0		5.8	30.3	K	8.3		56.9	33.0		9.5	8	5.0	54.6		8.3		19.8	7.6	
8.5		57.8	56.5	K	9.5		11.0	28.8		9.5	51	41.3	21.0		9.5		27.8	21.0		8.0		55.3	5.6	
7.8		58.5	35.7	L	9.3		14.3	56.9	K	7.5		51.3	13.8	K	9.1		31.0	10.1		9.3	20	5.8	53.0	
8.8	19	5.5	30.9		9.5		51.8	14.8		8.3	52	10.9	18.5	K	9.2		54.3	8.0		8.5		40.5	40.5	
9.3		31.8	48.1		9.5		54.3	35.3		9.5		15.3	27.6		9.5	9	10.5	20.2		9.1		17.4	58.8	
8.7		35.6	45.1	B	9.5	36	2.7	42.7		9.3		15.7	42.4		9.0		25.0	38.8	K	8.9		50.0	28.2	R
9.5		53.0	16.6		9.5		24.1	59.4		9.5		33.0	59.6		9.4		30.5	31.4	B	9.5		50.8	41.8	
9.5	20	4.3	0.1		7.7		37.0	18.3	L	9.5		48.0	47.8	K	9.5		35.0	25.6		9.5		53.2	20.8	
9.5		7.4	55.9		8.8		56.2	50.5		9.3		51.8	55.4		9.4		35.8	3.7		9.1	21	34.2	47.0	
9.5		9.8	5.7		9.4	37	0.9	36.2		8.3		53.5	48.1	K	9.3		46.1	24.7	B	8.8		34.5	18.1	K
9.5		19.0	32.0		9.2		22.7	13.7		9.4	53	5.3	3.1		8.7	10	8.3	26.8	K	6.5		53.6	15.6	K
9.1		22.5	46.4	K	8.2		36.6	0.4	K	8.5		25.7	39.9	K	9.3		18.6	8.0		9.5		59.8	26.2	
9.2		26.0	37.7	B	9.5		45.7	34.0		9.5		28.0	19.3		9.5	11	11.4	13.8	B	9.1	22	18.0	32.4	
9.3	21	6.0	36.6	K	9.0		50.0	55.4		9.3		39.9	54.8		9.5		23.8	34.3	B	9.4		27.8	38.3	
9.5		39.0	17.5		8.6	38	9.5	15.8	K	9.5		47.4	46.3	K	9.4		24.2	40.6		7.2		40.5	25.6	K
8.3		40.3	3.0		8.8		17.6	23.4		9.5		55.3	34.8		9.4		29.7	18.3		9.4	23	25.4	8.3	B
8.7		43.3	45.5	K	9.3		30.0	20.9	B	9.2		58.6	21.4	B	9.5		41.0	24.9		9.5		27.4	12.3	
9.3	22	9.5	53.5		9.4		35.1	14.8	B	7.6	54	1.5	59.3	K	9.5	12	33.0	20.4		9.5		28.0	49.2	K
9.4		11.1	58.5	B	9.0		38.7	19.1		9.4		9.9	17.5		9.3		58.3	54.1	B	9.5		39.2	43.5	
9.5		21.2	59.8		9.5		55.3	40.9		9.5		20.2	11.5	B	9.5	13	3.2	8.6		9.0		50.8	52.5	
8.5		21.7	34.1	B	9.0	39	12.7	23.0	K	8.5		23.0	38.8	K	9.3		10.4	59.1		9.0	24	14.5	40.3	
9.5		28.9	56.2		9.2		21.8	20.5	B	7.3	55	55.3	33.2	K	9.5		12.3	38.2		8.8		18.8	19.6	
9.5		41.2	3.9		9.5		29.5	35.1		6.5		59.9	0.9	K	8.8		39.2	1.1		8.8		22.4	29.9	R
9.4		58.6	22.9	B	9.4		30.4	3.7		9.5	56	1.7	37.9		7.3		55.8	40.7	K	9.2		23.8	2.1	B
8.9	23	32.7	51.8		9.3		40.6	19.8		9.5		15.8	45.1		9.1	14	11.7	35.4	K	9.5	25	13.4	55.3	
8.8	25	7.4	47.3	K	9.3		49.7	28.4		9.5		45.3	18.8		9.4		14.1	52.5	B	9.5		42.8	27.9	
9.5		27.6	2.7		9.5	40	38.8	26.6		9.1	57	23.0	52.4	K	8.5		19.8	5.1		8.9		28.5	14.0	
9.2		29.3	11.2	B	9.1		51.0	54.5		9.3		48.0	23.3		8.2		20.0	48.5	K	9.5		45.2	54.0	
9.3	26	11.5	11.6		8.3		59.9	13.6	K	9.1		58.6	14.4		9.5		29.1	56.3		9.5		58.1	7.8	
8.2		43.6	3.6	K	9.5	41	12.0	7.3		8.8	58	23.5	30.2	K	8.5		34.0	29.4		9.5	26	18.3	45.7	
9.0		50.5	8.2		9.0		13.0	41.4	K	9.2		33.7	55.3		8.5		36.8	12.5		9.4		42.7	30.1	
8.9	27	0.3	54.3		9.5		15.5	49.2		var		44.5	24.5	K	7.5	15	12.3	13.8	L	9.2		46.1	12.8	
8.6		23.0	50.2	K	9.5		51.6	7.7		9.1	59	31.3	4.6		9.4		23.8	18.3		9.1	27	19.3	37.7	
9.3		24.8	36.1		9.3	42	2.7	19.6		8.6		38.0	44.3	L	9.4		30.0	1.9		9.1		24.6	4.8	
9.0		28.0	27.6	K	9.0		51.5	4.7		8.6		40.0	16.3	K	6.5		34.0	44.3	K	8.0		45.9	2.7	K
9.0		38.3	37.7		9.1	43	6.2	4.2		9.5		58.4	9.0		7.0		45.7	27.0	L	9.1		56.3	5.3	
9.3		39.5	50.5		8.8	44	6.9	3.9		9.0	0	5.3	48.0		9.5		50.8	24.0		9.1	28	6.8	16.3	
9.5	28	11.6	52.4		9.3	46	27.0	49.5		9.2		14.4	55.1		9.5		53.2	57.5		9.5		15.2	5.6	
9.2		20.3	57.3	R	9.1		31.2	54.8		8.5		23.5	22.5	B	9.5		53.7	15.1		9.4		22.3	31.7	B

lower edge of the chart, a star designated as variable ("var") can be identified. This is Algol, a well-known variable in the constellation Perseus whose brightness varies within a period of a few hours. It has the catalogue-number + 40°673.

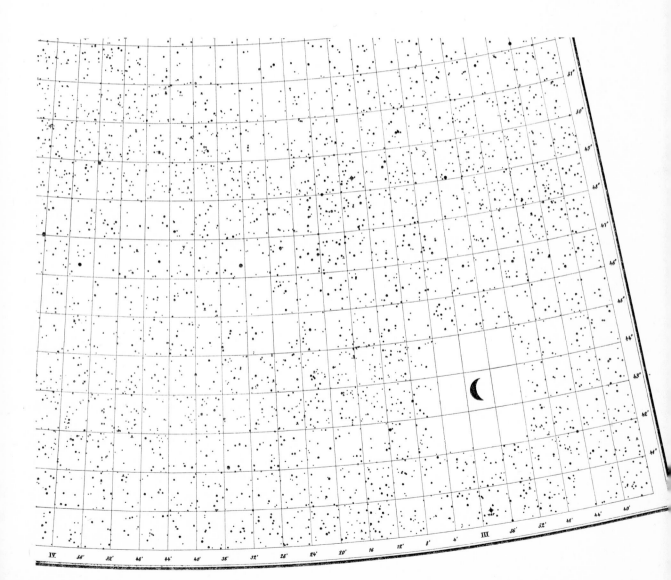

would be the result. With this size of instrument, a rigid construction in which position-dependent deformations are totally excluded would not be feasible simply on account of the enormous cost. This is why the decision was taken to accept deliberately such deformations but, in the supporting-system chosen, to allow for them in such a manner that the reflecting surface would have a parabolic shape in every position. The individual shapes obtained when the instrument is moved then have different focus positions, it is true, i. e., the rays are now focussed at somewhat different points above the reflector surface, but this can be compensated by shifting the receiving antenna. The elastic deformations lead to displacements of up to 75 mm at the edge of the reflector but the overall surface always remains a paraboloid, with deviations of less than 3 mm. The design calculations had to take account not only of gravity but also of the effect of wind pressure and thermal stresses caused by solar heating. Up to a diameter of 65 metres, the surface of the reflector s covered with special panels, 1 m $\times$ 3 m in area, up to a diameter of 85 metres there then follow reinforced aluminium panels and finally, on the outer edge there is wire-netting with a 6-mm mesh. The high surface accuracy permits the effective reception of radio waves down to 8 mm wavelength and thus meets the requirements of cosmic applications. At a wavelength of less than 3 cm, however, the telescope is used with a reduced aperture since the wire mesh around the outer edge allows the very short waves to pass through. The great "reflector bowl" is tilted in the vertical position by turning around a shaft which is supported by two towers at a height of 50 metres above ground-level. For obtaining a setting in the azimuth direction, these towers are moved around a large circular rail. All the movements of the telescope are controlled by a computer in accordance with programmed coordinates. This computer also brings the receiving aerial to the specific focus position, registers the measurement data and carries out the first data-reduction procedure. The azimuth direction of the telescope posed particular problems during its construction. On account of the accuracy of adjustment stipulated, the circular rail of 64 metres in diameter had to be a perfect circle to within 0.2 mm and, under the load of 3,200 tons represented by the moving weight of the telescope, was not to settle by more than 1 mm in an uneven manner. This was achieved by laying a ring of concrete four metres wide which was located in solid rock by 140 reinforced concrete supports of 1.2 m thickness and 7 to 10 metres in length.

Like optical reflectors, radio telescopes of the Effelsberg type can be designed for the direct detection of the radiation concentrated at the prime focus or the radiation can be reflected by a secondary mirror and only then pass to the actual feeder aerial for the receiver. The latter variation is better for wavelengths of less than 11 cm and it has the advantage that several feeder aerials for different directions or frequencies in close proximity can operate simultaneously and thus the radio sources can be scanned spatially or spectroscopically at the same time. For the accommodation of feeder aerials and receivers, the primary and secondary focus cabins are arranged 30 metres above the reflector and at its base. Since the radiation from cosmic bodies in the radio-frequency range is very weak, a high degree of sophistication is required both for the feeder aerials and the receivers.

The possibility of detecting radio waves between a few millimetres and a wavelength of 20 metres and using them for the exploration of Space has led to essentially new findings in all areas of astronomical research. The radiation sources investigated by the staff of the Max Planck Institute are either in extragalactic Space or they are objects in the Milky Way, such as the central region, areas of neutral or ionized hydrogen, dark clouds, pulsars or other sources or the planets. For the investigation of the radio-frequency radiation emitted by the Sun, they have a special 10-metre telescope at their disposal. The research carried out with

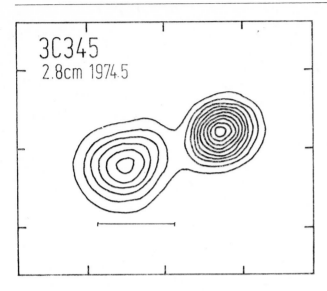

8   The lines of equal radio intensity (wavelength 2.8 cm) of the quasi-stellar radio source 3C345, mapped by the 100-metre radio telescope of the Max Planck Institute for Radio Astronomy in an intercontinental interferometer hook-up, demonstrate the unsurpassed resolving power of this observation technique. Optical astronomy designates these objects as "quasi stellar" whereas radio astronomy is still able to identify a fine structure. With a resolving power of this order, optical telescopes would show the image of the Moon with details of 1 metre in size. (Scale: $\frac{1}{1000}$ second of arc)

the 100-metre telescope includes both the continuum radiation and the line emissions or absorptions emitted by gaseous matter in Space. This includes the highly interesting subject of the investigation and discovery of interstellar molecules—some of which are of great complexity—which, for physical reasons, can be precisely observed in wavelengths of the centimetre range. The new 30-metre radio telescope which is to be built for measurements of up to 1 mm wavelength will also be used for this area of research. This installation is to be built in Southern Spain and may possibly be jointly operated with France.

There is one field of work which merits special mention. The angular resolving power in the reception of electromagnetic radiation is dependent on the ratio of the wavelength to the diameter of the aperture and is therefore very much worse in the radio range than in the optical one. This can be compensated by coupling several radio telescopes together in an interferometer arrangement. The individual components then function like parts of a very large telescope. The magnitude of such layouts extends to the VLB-interferometer (VLB = very large baseline) in which the aerials of different countries or even continents are harnessed together. Each individual aerial follows the source investigated for a fairly long time and the data received is recorded on magnetic tape. The tapes are sent to a common processing centre and after a combined evaluation procedure, a picture of the source is obtained with a high degree of resolution. The 100-metre radio telescope of the Max Planck Institute is a link in several such arrangements and networks and its most important partners are similar institutes in the Soviet Union, Sweden, United Kingdom and the USA.

● Konkoly Observatory of the Hungarian Academy of Sciences is situated on one of the elevations on the western outskirts of Budapest, 350 metres above the Danube. It takes its name from Dr Nicolaus Thege von Konkoly who founded a private observatory at Ogyalla in the second half of the last century. This observatory was transferred to the State in 1899 and it was ultimately relocated in Budapest in 1920.

The research programme of the Konkoly Observatory includes purely theoretical investigations, such as the application of the laws of plasma physics to astronomical problems, and also extends to activities dealing with the structure of our stellar system in which observations play a dominant role. The observation of artificial satellites is likewise represented at the observatory in Budapest or at its outstation at Baja in Southern Hungary. This sphere of activity will become even more important in future since laser techniques are to be used for measuring the precise distances of satellites. The theoretical projects follow in the great traditions of mathematics and physics in Hungary. However, the reputation of Konkoly Observatory among experts is based above all on the work carried out for many years on the exploration of variable stars. In fact, the stars are not so invariable as might be thought. There are very many stars which, over greater or lesser periods of time, periodically or irregularly change in luminosity. These bodies can be classified in different groups according to the shape of the light-curve. The interest of Hungarian astronomers has always been concentrated primarily on pulsating stars. The contributions of Konkoly Observatory in the field of variable star research have been acknowledged by the holding of international conferences on this subject in Budapest on several occasions.

It is also worth mentioning that an information bulletin of the International Astronomical Union is issued regularly by Konkoly Observatory, thus enabling new results from this branch of research to be published without delay.

At the Institute in Budapest, a 60-cm reflector telescope is used for the photoelectric monitoring of variable stars. However, since the early 1960's most of the observation work has been carried out at the mountain-station on Piszkéstetö in the Matra Mountains, this being at a distance of some 120 km from the capital. It is located on a plateau rather than a peak, there are no large communities in the vicinity and favourable conditions for astronomical observations therefore exist. The architecture of the buildings is unusual but the use of natural stone means that they blend very well with the local landscape. The two large telescopes here complement each other excellently in the facilities they offer the astronomers. The Schmidt camera, with its 90-cm mirror and 60-cm corrector plate, is used for photographical observations. On account of the large field of view in this system, it is particularly suitable for star-field photographs for the purposes of stellar statistics, for photometric work and for the detection of variable stars. The latter bodies also include the supernovae. These are stars in which a mighty increase in brightness suddenly occurs as an explosion in a late stage of their development. To come closer to the exact causes of this phenomenon, it is still necessary, at present, to collect material by regular photographic monitoring of as many galaxies as possible. The careful examination by the Budapest astronomers of numerous Schmidt plates exposed with this aim in mind has been rewarded on many occasions by the discovery of such bodies.

It is true that astronomical star-field photographs supply information about a large number of stars at the same time, but the study of these photographs has to be supplemented by the detailed investigation of individual bodies by photoelectric or spectroscopic means. Thus, for example, the photoelectric measurements of some stellar brightnesses must supply the calibration without which the images on photographic plates cannot be converted to the astronomical brightness system. This necessary supplementary information is supplied in the Matra Mountains since 1974 by a 1-metre telescope of the Ritchey-Chrétien type. It is equipped with: photoelectric photometers for measuring the intensity or polarization of stellar light; a spectrograph for determining the spectral composition of the radiation; and a photographic system which, although it has a smaller field of view than the Schmidt camera, has a greater focal length and thus superior resolution of stars which are very close to each other, as in star clusters, for instance. A modern digital system handles both the processing of the data and also a programmed, automatic movement of the telescope from one star to another. For this, a large computer is employed which is housed in the same building.

The development of the technical side of astronomy is particularly evident from the example of Konkoly Observatory. Only 20 years ago, it was the practice, with the 60-cm telescope referred to above, for an assistant observer in a room below the dome to take the readings from an indicator instrument and note down the measurements. In comparison with that, efficient photometers with instant data transfer to the computer and automatic telescope control are now at the disposal of the astronomers there. This ensures optimum use of the nights available for observation, which are relatively seldom in Central Europe, provides the astronomer with a survey of his results at all times and enables him to devote his full attention to important checking routines.

● The Armenian Soviet Socialist Republic lies in the latitude of 40° north and enjoys a mild and especially very dry climate. The cold air from the North is kept away by the Caucasus and the Black Sea in the West also plays its part in preventing the occurrence of cold winters in Armenia. Meteorological observations over many years at Yerevan, the capital of the Armenian SSR, show that the average annual sunshine amounts

to 2,711 hours, i. e., the sun is shining for almost two-thirds of the daytime. The annual rainfall is only 300 mm whereas in Moscow the annual precipitation is 533 mm, in Berlin 570 mm, in Rome 803 mm and in Zurich even 1,019 mm. These few meteorological facts alone indicate that Armenia offers excellent conditions for astronomical observations. As it happens, Byurakan Astrophysical Observatory was founded in 1946 on the initiative of V. Ambarzumian, the present director of the Observatory and the president of the Academy of the Armenian SSR. V. Ambarzumian was also the president of the International Astronomical Union from 1961 to 1964. Byurakan Observatory is an institute of the Academy of the Armenian SSR.

The Observatory is situated about 40 km to the north of Yerevan at a height of 1,500 m. To the north, the terrain rises upwards to Mount Aragatz, an extinct volcano, 4,200 m high. To the south and on the other side of the Turkish frontier which is not far away, there is Mount Ararat, 5,300 m high and always snow-capped at the peak. In 1948, an investigation of star clusters began with the aim of obtaining information about star formation, a 20-cm Schmidt camera being used for this purpose. A subject of great importance in this connection was the discovery, in the early years of the existence of this Observatory, of stellar associations. These are local accumulations of stars with the same physical characteristics and they represent the least compact form of star clusters. The associations were found during the investigation of stars of the spectral types O and B, these being high luminosity hot stars. The investigation of the movement of individual stars in the association showed that the stars might have been formed only an astronomically short time before in a very restricted volume of Space. An age of 1.5 million years was postulated for an association of O-stars in the constellation Perseus. Our Sun, on the other hand, is 5,000 million years old.

The equipment of the Observatory was completed by a Schmidt camera with a 53-cm mirror and 53-cm corrector plate and another one with a 130-cm mirror and a 1-metre corrector plate. The "Big Schmidt" has three objective prisms which give a dispersion of 28.5, 90 and 170 nm per millimetre in the blue spectral region. With the smallest dispersion of 170 nm per millimetre, it is still possible to reach bodies of the 17th magnitude. Up to now, the 1/1·3-metre Schmidt telescope has been mainly employed for the investigation of peculiar galaxies and flare stars. The best-known discovery with this telescope is that of the Markaryan galaxies which are now to be found on observation programmes all over the world. Markaryan galaxies are characterized by a very strong continuum in the ultraviolet. Two groups of these galaxies are now known: the first consists mainly of spiral galaxies with a bright and well-defined nucleus as the source of the continuous ultraviolet radiation. The second group comprises diffuse galaxies in which the ultraviolet emission sources are spread across the entire body. The members of the first group include numerous Seyfert galaxies which can be recognized in particular by their star-shaped nucleus and their emission line spectrum with very wide emission lines. However, there are also Markaryan galaxies with very sharp emission lines.

The low-intensity members of the second group have spectra which are similar to those of ionized hydrogen regions. Hence these bodies are also sometimes attributed to intergalactic regions of ionized hydrogen. But among the weak bodies of the second group there are very many irregular galaxies with a low metal content. It is probable that these galaxies were formed only in the last few hundred million years.

The "Smaller Schmidt" has been used for the investigation of the nearer galaxies with very special attention being paid to the degree of concentration of the nuclei of the galaxies. The starting-point was the work carried out by V. Ambarzumian in 1958. He postulated that the nuclei of galaxies were of particular importance for the development of galaxies. To ascertain the degree of concentration, photographs with different exposure times were taken and the angular diameters of the nuclei measured. So far a total of more than 500 spiral galaxies has been classified.

The "Smallest Schmidt" was and still is used in conjunction with a 50-cm and a 40-cm Cassegrain reflector for the investigation of flare stars. These are young and unstable stars. The Schmidt telescope is employed for the monitoring of star clusters and associations so that flare phenomena can be detected. The Cassegrain telescopes are equipped with sensitive photometers and polarimetric devices for the observation of known flare stars.

For reduction purposes, the astronomers at Byurakan have a two-dimensional digitalized isodensitometer, microphotometers, blink comparators and a computer at their disposal. In 1976, the equipment of the Observatory was again upgraded to a significant extent. On the model of the 2.6-metre telescope of the Crimean Observatory, a 2.6-metre reflector was also constructed for the Byurakan Astrophysical Observatory and was officially taken into service on 4 October 1976. For the Cassegrain focus, there is a powerful spectrograph available, permitting dispersions between 3.5 and 40 nm per millimetre. Electronic cameras of the Lallemand type enhance the performance of the new system.

Like many astronomers throughout the world, the 25 members of the scientific staff of Byurakan Observatory not only carry on research but are also active in teaching and education. The scientists from Byurakan hold lectures at Yerevan University and act as tutors for students from this university who are writing their theses at the Observatory.

10   The dome of the 1.25-metre telescope of Abastumani Observatory is 12 metres in diameter and tops a three-storey building with broad terraces on each floor. The building houses the library of the Institute, a conference room for 100 persons, smaller seminar rooms and research offices.

11   So that a uniformly sharp image can be obtained of sections of the sky of several degrees in diameter with the aid of a reflector, optical corrector-members must be arranged in the light-path in addition to the principal mirror. The design by the Soviet optical worker D. D. Maksutov specifies a spherical principal mirror and a special lens, a meniscus, arranged near to the prime focus. The meniscus telescope of Abastumani Observatory has a 70-cm aperture and is the most powerful of its kind at present.

12   The archives of Abastumani Observatory contain tens of thousands of special photographs of the Sun and thus document the importance of this area of activity in the research programme of the Institute. This coronagraph is used for monitoring the corona of the Sun. The 50-cm objective is mounted at the upper end of the steel structure. The image of the Sun obtained in the focal plane is covered by a stop and only the light emitted by the corona is investigated. With the great difference in brightness between the disc of the Sun and its envelope, an atmosphere with an extremely low degree of scattered light above the observation site is essential for the acquisition of scientifically useful exposures.

13   The 1.82-metre telescope
is used for direct photography
with an aperture ratio of 1:9,
i. e., the focal length in this case
is 16.4 metres. The range limit
for direct photography is the
22nd magnitude. The dome for
the telescope is 16 metres in
diameter and sufficient space
is consequently available for
the installation of equipment
for observation at the prime
focus.

14 F. S. Archenhold (1861–
1939) studied astronomy from
1882 to 1889 at Strasbourg and
Berlin. After his studies, he
worked for a short time at the
Urania Observatory in Berlin.
From 1890 to 1895, he was no-
tably active in the field of celestial photography at Berlin Observatory and played a major
part in the establishment of a
photographic outstation of
Berlin Observatory at Grune-
wald Forest. It was due to his
initiative in particular that the
observatory in Berlin-Treptow
was built which now bears his
name.

15 The 21 metre long "sky
cannon" of the Archenhold
Observatory is now one of the
historical features among the
astronomical instruments on
show. It stands on a foundation of 60,000 bricks and is always the centre of attraction
for many thousands of visitors.

Page 46 (Archenhold Observatory):

16 The small planetarium of the Archenhold Observatory is an important aid in the dissemination of astronomical knowledge. In addition to the public demonstrations, it is used by the staff of the Observatory to supplement the instruction given in astronomy at the polytechnical secondary schools of Berlin.

17 The dome building of the 2.6-metre telescope is built in a very unusual style. The entrance tower is located a few metres away from the building to which it is connected by a bridge. In the picture, the entrance tower is in front of the dome building.

18 The 2.6-metre telescope has an equatorial fork mounting and a skeleton tube. The instrument can be used with an aperture ratio of 1:3.6 at the prime focus and also has a 1:16 Cassegrain focus and a 1:40 coudé focus.

19   The illustrious classical architect K. F. Schinkel had a major hand in the design of Bonn Observatory which was constructed between 1839 and 1845. This is why the building is now scheduled as an Ancient Monument. The observations for the *Bonner Durchmusterung* were made in one of the six smaller towers at the side, one of which can be seen at the edge of the picture.

20   F. W. Argelander's 7.5-cm telescope was described in 1913 by E. C. Pickering, the director of Harvard College Observatory at Cambridge, Mass., for many years, as the world's smallest telescope with which the greatest task was performed. This was a reference to the *Bonner Durchmusterung*, a catalogue listing 324,000 stars. The instrument was built in the workshop of J. von Fraunhofer.

21   The "brain" of the radio telescope of the Max Planck Institute for Radio Astronomy at Effelsberg is the control room. The movements of the reflector are initiated and monitored from the left-hand side of the control console while the right-hand part is linked with the detectors for the radiation components in the millimetre to the decimetre range and monitors the observation process.

Page 50:
22   The model of the 100-metre radio telescope of the Max Planck Institute for Radio Astronomy as calculated by the computer showed that the shape of an upturned umbrella complied best with stability requirements. 24 of the radial struts seen in the picture support the reflector, which is constructed as a double-shelled "dish", in relation to the counterweight visible at the lower edge of the picture. At the time that the photograph was taken, only a small, inner part of the reflector had been installed. Some of the cover panels, which will eventually make up the coherent surface, are visible. The massive dimensions of this technical wonder can be estimated by a comparison with the stairs of the azimuth tower, on the left of the picture.

23   At the dome building of
the 1-metre telescope of the
Konkoly Observatory in the
Matra Mountains, all possible
precautions have been taken
to avoid disturbances of the
"astroclimate". Its height above
the ground greatly reduces the
troublesome influence of layers
of air just above the ground
while the aluminium skin, the
double walls and the layers of
insulation prevent a heat build-
up from outside in the dome
during the day. The ancillary
building, which is separate from

the dome, has a grass-covered
roof and cannot therefore be
a source of troublesome air
turbulence as would be the
case with a concrete roof exposed to strong sunlight.

Page 52:
24   Comparisons of astro-
photographs taken at different
times may reveal, in certain
circumstances, changes in the
position or brightness of the
objects photographed. Use of
this finding is made in the
search for specific phenomena.
In the present case, this comparison was obtained by super-
imposing two photographs
and interposing colour filters.
A supernova thus appears as a
striking red object. This supernova flared up in the outer part

of the galaxy in 1967 and
was observed for months as its
brightness slowly declined.
The picture was assembled from
plates taken at the University
Observatory, Jena, with a duplicate instrument of the 60/
90-cm Schmidt camera of Konkoly Observatory in Hungary.

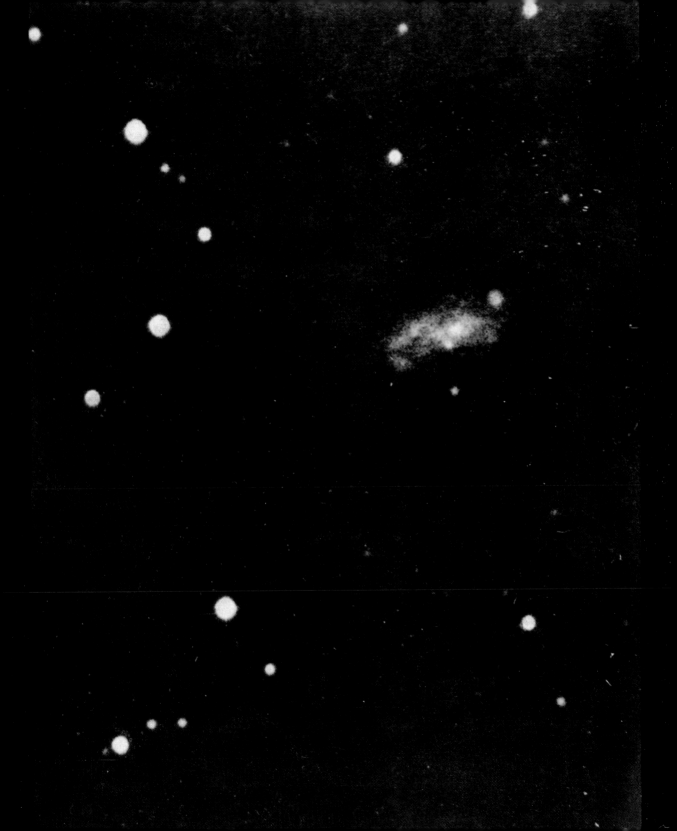

Cambridge, Massachusetts, the famous university city close to Boston, is the seat of a research institute which owes its present-day eminence to the fusion of two astronomical institutions possessing fine traditions.

In 1973, the Smithsonian Astrophysical Observatory and the Harvard College Observatory officially joined forces and became the Center for Astrophysics at which the work of almost 150 scientists is now coordinated. The formal foundation of this institution had been preceded by almost twenty years of close scientific collaboration.

● In his last will and testament, the property of the Englishman J. Smithson was bequeathed to the USA in 1836. The bequest, worth more than 500,000 dollars, was made on condition that an institute should be established under the name of the "Smithsonian Institution" at Washington to disseminate and spread scientific knowledge. This vague formulation led to many debates concerning the use of this great sum and even the President and Congress were involved in the affair. The plan to set up an astronomical observatory was stubbornly advocated from the very beginning but it was only in 1890 that it was actually implemented by S. P. Langley, the secretary of the Smithsonian Institution. S. P. Langley, an outstanding astronomer of the time, built this observatory so that he could continue his own observations of the Sun. In this, however, he acted in accordance with the views of his predecessors who believed that the best way to comply with the will of J. Smithson was to found an astrophysical research centre. In contrast to the tasks of classical astronomy, the aim of the work to be carried out was ". . . to investigate the nature and changes of the constitution of the heavenly bodies; to study the various emanations from these . . . and . . . to record and investigate the different phenomena which are included under the general term of terrestrial physics." (J. Henry, Secretary of the Smithsonian Institution, 1870)

This far-sighted goal and the scientific stature of S. P. Langley left their mark on the research profile of the Smithsonian Astrophysical Observatory. In the decades of its existence, the main fields of activity have been the investigation of solar radiation in its various guises and problems of the terrestrial atmosphere and the effect of the Sun on the Earth. From the very beginning, this also included important improvements and developments in equipment. As long ago as 1917, collaboration began with R. H. Goddard, the American pioneer of rocket technology, with the aim of using high-altitude rockets for solving the research problems of the Observatory.

Without waiting for the end of protracted negotiations concerning a possible site for an observatory, S. P. Langley had a wooden structure erected within a period of only five months in 1890 on the site of the Smithsonian Institution in Washington. This provided accommodation for the instruments and enabled the first observations to be made. This building was initially very simple and had only been planned as a temporary structure but with only a few changes and extensions it remained for many decades the home of the Smithsonian Astrophysical Observatory. However, so that the influence of the terrestrial atmosphere on the solar radiation measurements could be kept as low as possible, a series of outstations were established at high elevations in the course of the years. These included sites in California and Arizona while use was also made of the astronomically outstanding climate of Chile and of well-situated observation points in South-West Africa and Egypt.

Great changes had to be made to keep pace with the rapid development of astronomy after the Second World War. These included the transfer of the seat of the Institute from Washington to Cambridge, Massachusetts, and the link-up with Harvard College Observatory.

● It is true that this observatory was founded at Harvard University in Cambridge, Massachusetts, as early as 1839 but it was almost ten years later that its real development began. Advantage was taken of the great public interest aroused by the comet of 1843 to obtain substantial donations for the expansion of astronomical research at Cambridge. The sum raised permitted the construction of the University's own observatory building and the purchase of the Big Refractor from Merz & Mahler of Munich, the firm which continued the work of J. von Fraunhofer. The refractor was of advanced design and, with its aperture of 40 cm, is justly described as a big refractor for that time. This telescope is still in existence as a museum exhibit.

The historical merits of Harvard College Observatory consist in the pioneering work carried out in the introduction of photography as a new technique of observation in the second half of the last century, in its application for photometric purposes and in the acquisition of a great deal of spectroscopic material. When the word "photometry" is mentioned in connection with Harvard College Observatory, one immediately recalls the discovery by Miss H. S. Leavitt in 1904 of the relation between the light-change period and luminosity in variable stars of the Delta-Cephei type. For many decades, this relation has represented the basis for the determination of the distance of extragalactic star systems. Modern spectroscopy is based in two respects on work carried out at Cambridge. It was there, in the years before 1901, that the classification procedure for star spectra was developed which, in principle, is still used today and also in about 1920, after six years of work, a catalogue was produced listing the spectral classes of 225,000 stars. In addition to E. C. Pickering, the director of the Observatory at the time, credit must be given to a number of women astronomers who played an important part in this enormous project. This survey was made possible through a financial trust established by the widow of H. Draper, a man who made an important contribution to astrophotography,

and the spectral catalogue bears his name. By virtue of the mass of material listed in the *Henry Draper Catalog*, it has been used for important statistical investigations of great significance for our understanding of the stellar system in particular.

No mention has been made so far of the outstanding contributions of H. Shapley, one of the great astronomers of our age and director of Harvard College Observatory from 1921 to 1952. He calibrated the period-luminosity relation and made it a tool for the determination of distance in Space. Its application to the globular star clusters surrounding our Galaxy led to the bold conclusion that the Sun is located a long way away from the centre of this system and not, as previously assumed, in the midst of a very much smaller configuration. Other important works by H. Shapley are concerned with the extragalactic star system, galaxies as he called them.

During the time that the Observatory was headed by H. Shapley, the number of instruments available increased from about ten to thirty. Before 1921, the largest instrument had an aperture of 60 cm but by the end of the 1930's a 150-cm telescope was in service. Mention must also be made of special photographic cameras, instruments for the observation of meteors, equipment for solar research and a radio telescope. To the outstations which were already in use but were subsequently abandoned again, there was added in 1932 the Agassiz station. This is not far from the town of Harvard, Massachusetts, and subsequently became the most important observation centre of the Observatory. Since 1952, it has been known by the name of the man who made such a great contribution to astronomy at Harvard University.

Photographic monitoring programmes carried out with the aim of discovering and investigating variable stars have formed part of the work of this Observatory since 1898. Special cameras making a sweep of the visible sky twice a month have supplied so far a total of about 100,000 photographic plates which account for a considerable part of the material in the archives of the Observatory.

Another event of importance was the founding by E. C. Pickering of the "American Association of Variable Star Observers" and due mention must be made of the active support always given by Harvard College Observatory to this organization which now spans the world. Apart from the encouragement given to enthusiastic amateurs by professional astronomers by this, amateurs have traditionally contributed much valuable material to the understanding of the variable stars.

● The joint research potential of Harvard College Observatory and the Smithsonian Astrophysical Observatory permits the examination of a large number of astrophysical problems and the use of the most effective resources and methods. The field of activity ranges from the investigation of extragalactic star systems to planetary research, the latter subject including not only the major planets and the Earth but also the small bodies in the planetary system. In the field of solar physics, the classical results of solar observation are being linked and interpreted with modern data within the framework of the "Langley-Abbot Program". The name of the programme is a tribute to the pioneering work of two former directors of the Smithsonian Observatory. The research work in all fields is supported by contributions from areas of activity which are concerned with special questions of theoretical astrophysics or which measure or calculate fundamental quantities from nuclear or molecular physics.

There are two great reflecting telescopes of the 1.5-metre class at the disposal of the Center of Astrophysics. One of these is at the Agassiz outstation near Harvard, Massachusetts, while the other is located on Mount Hopkins, a mountain-station at an altitude of 2,600 m in Arizona. However, the spectral range under investigation includes not only the optically visible and infrared radiation, as picked up by typical astronomical telescopes, but also the radio waves emitted by cosmic bodies and X-ray and gamma radiation. Two radio telescopes, at the Agassiz station and at Fort Davis, Texas, and the 10-metre gamma radiation telescope on Mount Hopkins are used for these wavelengths. In addition to Earth-bound observation, measurements taken by instruments on board of satellites also play an important role. In fact, the Center for Astrophysics occupies a key position in the X-ray satellite programme of the USA.

The Multiple Mirror Telescope (MMT) on Mount Hopkins represents an innovation of great significance for the future in the field of astronomical engineering. Six separate mirrors, each 1.8 metre in diameter, concentrate the radiation picked up by them at a common focal point. In the way it functions, the system thus corresponds to a 5-metre telescope but the costs of manufacture are said to be very much lower. However, there is a complication in that the optical axes of the six sub-systems must always be aligned in such a manner as to be free from position-dependent flexing-effects, so that their individual images are accurately superimposed and thus give a sharp image of the object. This can only be achieved when the individual mirrors are not rigidly mounted in a frame but, on the contrary, are so arranged that they can be tilted by tiny amounts relative to the mounting with the aid of a complicated control system. Experience will show whether the instrument can justify the hopes and expectations placed in it. At any event, this concept represents an interesting attempt to avoid the enormous technical difficulties involved in the construction of telescopes with a mirror diameter very much greater than five metres.

● Mount Stromlo Observatory, Siding Spring Observatory and Anglo-Australian Observatory, Canberra/Coonabarabran, Australia

56

● It was on 16 October 1974 that the Prince of Wales inaugurated the Anglo-Australian 3.9-metre telescope of the observatory at Siding Spring in the South-East of Australia, about 450 km north-west of Sydney. The first proposal for the construction of this telescope was made as long ago as 1960. Four years later, negotiations began between the Royal Society and the Australian Academy of Sciences, which led to an agreement in 1967 between the Governments of the United Kingdom and Australia on the joint construction of a large telescope and its erection in Australia. The decision to site this telescope in the Southern Hemisphere followed from the wish to obtain more observational material on the Southern Sky which, at this time, had still not been explored to any great extent. Until the inauguration of the Anglo-Australian Telescope, the twelve largest instruments were all in the Northern Hemisphere.

The site selected is 1,165 metres above sea level. Long-term observations here have confirmed that Siding Spring possesses excellent atmospheric conditions for astronomical observations, i.e., the "seeing" is usually very good. Good "seeing" means that, due to low air turbulence, there is very little fluctuation in the brightness and direction of stars and that there is hardly any "twinkling". It is expected that 1,500 hours will be available every year for photometric observations at Siding Spring. An important factor in the selection of the site was the concentration of other important astronomical research centres in this area. At 280 km to the south-west of Siding Spring, there is the great Parkes radio telescope, while less than 180 km to the north at Narrabri there is the famous Hanbury Brown radio interferometer for measuring the diameters of stars, and the circular array of radio telescopes designed by Paul Wild for the investigation of the active corona of the Sun at Culgoora.

The nearest community in the vicinity of Siding Spring is the little town of Coonabarabran with 3,000 inhabitants, 33 km away. It offers the scientists and technicians working at Siding Spring all the comforts of civilization. Coonabarabran is easily reached from Sydney by air and there is also rail and road access. On the other hand, with only 3,000 inhabitants, the disturbances caused by civilization as regards astronomical observations such as glare at night time and the development of dust are still slight.

Siding Spring Observatory was originally built as an outstation of Mount Stromlo Observatory near Canberra which still remains the principal astronomical institution of the Australian National University. The history of the Observatory began in 1909 with founding the Australian Committee for Solar Physics. A 15-cm refractor which had formerly belonged to Lord Farnham, an English amateur astronomer, and a 23-cm refractor from another amateur, J. Oddie, were the first instruments of the new institution. The "Oddie", as the 23-cm telescope is known, was used for the first test-measurements on Mount Stromlo and is still in use today. On the basis of these test-measurements, a decision was taken by the committee in 1914 to build a great observatory there. However, as a consequence of the First World War, it was only on 1 January 1924 that the Mount Stromlo Observatory officially came into existence with W. G. Duffield as its first director and it is after him that the principal building of the Observatory is named. For its first two years, the Institute had to be accommodated in the Canberra Hotel, the new buildings of the Institute still being under construction at that time. In the 1940's, the Institution changed its direction of research, work now being centred on stellar astronomy instead of solar studies from this time onwards.

When Melbourne Observatory was closed in 1945, Mount Stromlo Observatory acquired the great 130-cm telescope for a price of only £ 300. This instrument had been built in Dublin in 1868 for Melbourne Observatory. Since it had a metal mirror and a tube of 24 metres in length, little use had been made of it. At Mount Stromlo, it was given a Pyrex mirror and, on account of the new optical system, the tube was reduced to

7.5 metres in length. The telescope was also equipped with an electrical drive. The reconstruction of the instrument enabled fundamental observations to be made of the Southern Sky. These included the first photoelectric observations of stars in globular clusters and in the Magellanic Clouds and the first infrared observations in the Southern Sky. In 1953, Mount Stromlo Observatory acquired a 1.9-metre reflector which, up to 1974, was the largest instrument in the Southern Hemisphere. It is mostly used for spectroscopic observations.

Since the observation conditions in the vicinity of Canberra were constantly deteriorating, a search was begun, lasting ten years, for a suitable site for a new observing station in Western and Southern Australia, Victoria and New South Wales. As a result of these investigations, work began on the new observatory at Siding Spring in New South Wales in 1962. Since then, this station has developed into an institution of international importance. In addition to the 1-metre, 60-cm and 40-cm telescopes of the Australian National University, it also accommodates, since 1973, the 1.2-metre Schmidt camera of the United Kingdom and the Anglo-Australian Telescope. This instrument is equipped with a glass-ceramic Cervit mirror. The thermal expansion coefficient of Cervit is almost 100 times less than that of glass so that there is practically no distortion of the mirror as a consequence of fluctuations in temperature. The surface accuracy of the Ritchey-Chrétien mirror is so great that in comparison the height of the Television Tower in Berlin (365 metres) would have to agree with that planned to within 0.01 mm. The ground Cervit mirror has an aluminium reflective coating of 120 nm in thickness. The application of this vaporized aluminium coating was carried out on site at the Observatory. Curiously enough, because of this, the first pictures with the telescope were obtained with the "naked" mirror before the mirror was coated.

With good seeing, the 23rd magnitude has been attained photographically in the blue spectral region with the finished instrument while stars of a magnitude of 22.5 have been observed in the visual spectral region. In good seeing conditions and with the aid of a very sensitive television camera, it has even been possible to show stars of a magnitude of 22.5 after an integration period of 15 seconds in the Cassegrain focus. It is a noteworthy fact in this connection that a magnitude of 22.5 can be reached within a relatively short time with the television technique and that problems essentially relate to the storing of the television picture. In the simplest case, this is even possible by photography but the field of view in the Cassegrain focus is smaller than in the prime focus which is used for direct photography.

The Anglo-Australian Telescope is one of the first and most comprehensively electronically controlled instruments. With this telescope, the observer no longer needs to be always in the dark and often cold dome so that, with his eye on the eyepiece, he can check and if necessary correct the tracking of the telescope. Instead he controls the instrument from a control centre or, more accurately, supervises a computer which controls the telescope and through which it is possible to program the telescope. With the Interdata 70 computer, orders can be given to the telescope manually, by typewriter keyboard, by punched card or punched tape. When the entire observation programme, with the coordinates of the different bodies to be observed, the filter data for photometry, the integration periods, the sequence of the object in the programme, and so on, is put into the computer, it takes over all the control functions. The coordinates fed to the computer do not need to be the true coordinates for the time of observation since the necessary precession, nutation, aberration and refraction corrections are calculated and included by the computer in an independent manner. It also handles the tracking of the telescope when observations are in progress. The aim is to track through the computer with an accuracy of 0.1 seconds of arc when seeing conditions permit this. For the checking of the entire system, there is a visual indicator instrument at the disposal of the astronomer on duty in the control centre and a keyboard there enables him to correct very rapidly and easily the course of the programme. A second Interdata 70 is provided for the acceptance of observational data from the TV camera, photometer, image intensifier or other electronic detectors with a digital data output. This computer also compacts the incoming data material and stores it on magnetic tapes which can then be evaluated at another time on other computers. This can be done, for example, at Epping near Sydney where the headquarters of the Anglo-Australian Observatory, together with the mechanical and electronics centre, the library, a computer installation and the administration are located.

Of the observation time available with the Anglo-Australian Telescope, 45 % is at the disposal of the Australian astronomers and 45 % is for the British staff. The other 10 % is reserved for the director. It is needed for testing and checking requirements in connection with the telescope, for testing new equipment and for the observation of unexpected celestial occurrences.

● At the end of 1972, the Royal Observatory Edinburgh celebrated its 150th anniversary with, among other things, an exhibition of its more recent scientific work. The most impressive exhibits were the fully automatic measuring equipment for photographic plates, results of observation with the European TD1 satellite which had just been launched at that time and equipment for astronomical infrared photometry. When it is considered that the Scottish Royal Observatory was given the responsibility for the national 1.2-metre Schmidt camera and 3.8-metre infrared telescope projects by the U. K. Science Research Council, a good idea of the scientific importance of the establishment can be obtained.

The history of Edinburgh Observatory as a royal observatory began in 1822 when it received this privilege on the occasion of a visit to Scotland by George IV. It was only a few years prior to this visit that an observatory had been built on a hill in the city-centre on the initiative of prominent citizens. This Carlton Hill Observatory is still in use today as a municipal observatory. At the end of the last century, the difficulties for astronomical work caused by the urban surroundings and probably the lack of space as well led to the erection of a new observatory complex on the southern outskirts of the city on Blackford Hill.

In connection with the history of the Observatory, mention must be made of the famous Crawford Collection in the library. Towards the end of the previous century, the Earl of Crawford presented the British Government not only with the instruments of his private observatory but, in particular, also with his collection, amounting to more than 10,000 items, of early astronomical books and manuscripts. By this he not only saved the Observatory from the threat of closure but also contributed to the construction of the new centre on Blackford Hill. The collection, an acknowledged treasure-chest for those studying the history of science, dates back to the 13th century and contains first editions of all the major works on astronomy and related subjects, including the writings of Copernicus, Apianus, Kepler, Galileo and, of course, Newton.

For astronomical observations, the Observatory has at its disposal a number of large instruments at Edinburgh, comprising a 50-cm and a 90-cm reflector, a twin reflector with automatic operation and a Schmidt camera with a mirror diameter of 60 cm. There is a duplicate of the latter instrument on Monte Porzio in the vicinity of Rome, together with a Michelson interferometer which is used for measuring the distances of double stars. This outstation was established because there were better climatic conditions there and accordingly more favourable working conditions than in Scotland. Overseas stations of other British astronomical institutes are maintained in the Canary Islands, in South Africa and Australia and on Hawaii.

For some little time now, British astronomers have been able to use the biggest telescope built for measurements in the infrared spectral area, this instrument being situated on the volcanic cone of Mauna Kea at a height of 4,200 metres on Hawaii. The location is ideal for work with wavelengths of more than $1\,\mu$ since at such an altitude there is scarcely any atmospheric water vapour present which, at a lower location, would absorb much of the long-wave radiation. With the development of sensitive detectors, the importance of infrared astronomy has enormously increased in the last decade. Eight different research teams are working on specific problems in this sphere in the United Kingdom. The 3.8-metre telescope will help them to make further progress in solving them. Even during the constructional phase, the scientists and technicians of the Observatory were entrusted with weighty tasks and now they bear the entire responsibility for the operation of the telescope. Before the "United Kingdom Infrared Telescope" was constructed, the possibilities and necessity of a larger instrument were demonstrated by a pilot telescope with an aperture of 1.5 metres on Tenerife, in the Canary Islands.

Depending on the size of the field of view, the aperture and the specific exposure conditions, photographs of star fields taken with modern telescopes can show up to several hundred thousand objects, mainly stars and galaxies. From the blackening on the plate, astronomers can work out exact positions in the sky, the brightnesses of stars or the brightness distribution in the pictures of the stellar systems. However, with conventional methods of evaluation, there is a very great discrepancy between the time needed at the telescope—about one hour per exposure—and the days or weeks needed for the evaluation. Despite the rationalization of the methods, the use of a large number of personnel and the restriction of the objects measured to a minimum, there is a danger that telescopes become blocked by a growing stack of exposures which have not been evaluated. This was a special problem for the Royal Observatory since the British astronomers had a particularly powerful telescope at their disposal from 1973 onwards in the form of the 1.2-metre Schmidt camera at the Anglo-Australian Observatory at Siding Spring in Australia. In collaboration with industry, two types of fully automatic measuring equipment were developed which determine and prepare all the data desired from the photographs to a high degree of accuracy and without human assistance. The designations of the systems—GALAXY and COSMOS—would appear to come from the astronomical vocabulary but in fact these are acronyms. COSMOS is made up of the initial letters of the words "coordinates", "size", "magnitude", "orientation" and "shape", indicating the ability of the apparatus to distinguish stars from galaxies and to deduce characteristic measurement data. Other programmes concern the investigation of stellar spectra or diffuse cosmic objects and they can also be used for non-astronomical applications, provision being made for this. The working-speed is very high and equals at least 1,000 objects per hour and, if the degree of accuracy is restricted, can even be very much higher. With a measuring process of such rapidity, the direct evaluation of the plates no longer represents a problem. The astronomers now have to make a rapid scientific interpretation of the data material obtained. An effective organization of work is the basic condition for this.

● In 1982, the McDonald Observatory will be able to look back on a history of 50 years. The basis for the construction of the Observatory was a trust established by the Texan banker and amateur astronomer W. J. McDonald in 1926. His will specified that the University of Texas could use the money only for the construction of an observatory. The site of its construction was not specified in the will and, after a great deal of thought had been given to the matter, it was decided to build it on a mountain of 2,100 metres in height located at about 27 km from Fort Davis. The farmers who owned the land donated it to the University for the construction of the Observatory. This is still recalled by the name of the mountain, Mount Locke, on which the Observatory stands. It was named after Violet Locke McIvor, the wife of a farmer.

The site selected for McDonald Observatory was an excellent choice, especially in regard to the industrial development which has had such negative effects on astronomical observations in many countries of the world. McDonald Observatory is far removed from the great cities and industrial complexes. El Paso is more than 300 km away and Austin, the seat of the University of Texas, is at a distance of some 800 km.

From 1932 to 1962, McDonald Observatory was jointly run by the University of Texas and the University of Chicago but since 1962 it has been the sole responsibility of the former. The first instrument of the Observatory on Mount Locke was taken into service on 5 May 1939. This 2.1-metre reflector is now known as the Otto Struve Telescope. O. Struve, the last of the long line of Struve astronomers, was the first director of McDonald Observatory. It was only in 1956 that the Institute acquired a second instrument in the shape of a 91-cm reflector, this then being followed in 1969 by the 2.7-metre reflector, and in 1970 by a 76-cm Cassegrain instrument. In its almost 50-year's history, McDonald Observatory can look back on numerous scientific achievements. It was with the telescopes of the Observatory that G. P. Kuiper discovered a satellite of Uranus and a satellite of Neptune. O. Struve demonstrated the existence of hydrogen atoms in interstellar space. It was at McDonald Observatory that interstellar polarization was observed for the first time. Evidence was found for the seasonal variation of water vapour on Mars and molecular oxygen was discovered in the atmosphere of Mars.

The discoveries of the past show that a broad spectrum of astronomical research has been pursued at McDonald Observatory. This still remains true of the present. Spectroscopic investigations of the planets are being made, the movement of the Moon is being accurately observed with the aid of laser beam reflections, the abundances of the chemical elements in evolved stars is being studied, photometry with a high time resolution of variable stars is being carried out and studies are being made of the evolution of galaxies.

In addition to the optical telescopes at their disposal, the hundred scientific workers at McDonald Observatory also have a radio telescope for observations in the millimetre range.

25 The 1.5-metre reflector of
Harvard College Observatory
was built in 1933, a new mirror being installed four years
later. The telescope is still in
use today at the Agassiz Station
and remains a powerful instrument.

26   This fully steerable radio telescope with a reflector of 25 metres in diameter is at the Agassiz Station of Harvard College Observatory and is shared with the Astrophysical Institute of the Smithsonian Institution.

27   The photograph shows the spectra of numerous stars taken with the aid of an objective prism. The individual spectra can be assigned to specific classes on the basis of the dark absorption lines. This alone enables rough conclusions to be drawn about the physics of stellar matter in the outer, light-emitting layers. Photographic plates of this kind were also used for the production of the *Henry Draper Catalog.* This picture shows the area around the star Eta Carinae with its emission nebula and was taken in 1893 at the Boyden Station (Harvard College Observatory) which was still at Arequipa (Peru) at that time.

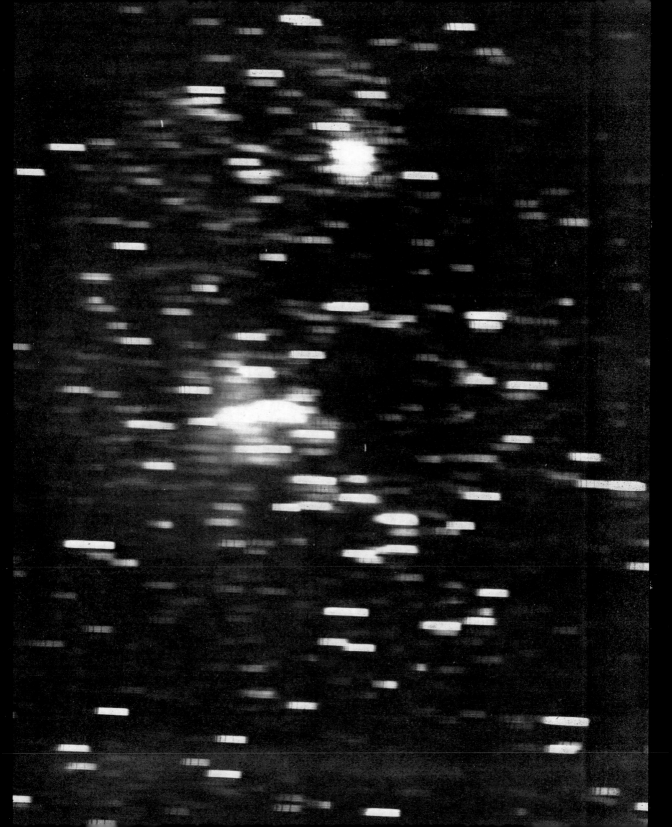

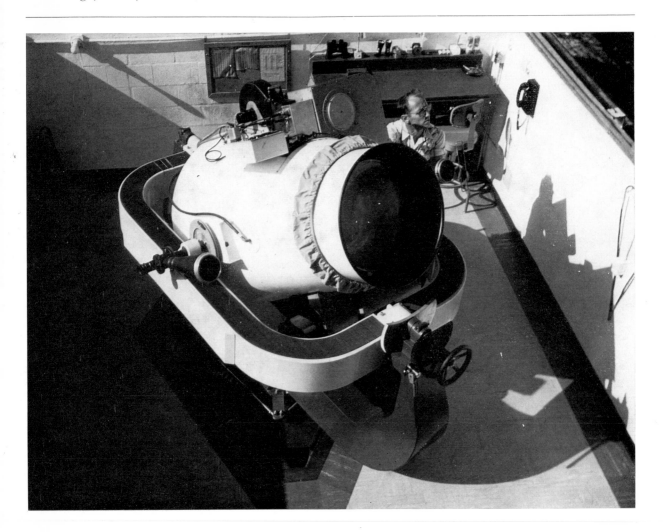

28　Covering every continent, a worldwide network of wide-angle photographic cameras, known as Baker-Nunn cameras, are used for the determination of the exact positions of artificial terrestrial satellites. From the data acquired, conclusions can be drawn as to the shape of the Earth, the gravitational field and the terrestrial atmosphere.

29　Since 1966, the mountain-station of the Smithsonian Astrophysical Observatory has been located on a ridge of Mount Hopkins, 2,600 metres above sea level. The altitude, the dry climate and the large number of clear nights in Southern Arizona provide exceptionally favourable conditions for astronomical observations.

30　The Multiple Mirror Telescope of the Harvard Smithsonian Center for Astrophysics on Mount Hopkins features a number of mirrors in a supporting skeleton and with a common focus. The structure housing the telescope also represents a new concept: it encloses the telescope in a space-saving manner and rotates with the telescope.

Page 66:

31　High-energy gamma rays entering the Earth's atmosphere produce individual flashes of light which are gathered with this telescope-like mosaic light collector and passed to a sensitive radiation detector. The low intensity of the phenomena to be detected necessitates exceptional observation conditions.

Mount Stromlo Observatory, Siding Spring Observatory
and Anglo-Australian Observatory, Canberra/Coonabarabran, Australia

67

32 At Siding Spring Observatory, there are other instruments besides the Anglo-
Australian Telescope. In the
foreground of the picture, there
is the telescope building while
at the rear there is a good view
of the dome of the 1.2-metre
Schmidt telescope. This Schmidt
camera and the 1.2-metre
Schmidt telescope of the European Southern Observatory at
La Silla in Chile are being used
to supply the photographs for
the photographic atlas of the
Southern Sky.

33 The prime focus cabin of the telescope is moved with the instrument. The focal length at the prime focus is 12.9 metres, i. e., the observer sits about 13 metres in front of the primary mirror. The prime focus is mainly used for direct photography. The field photographed is 1 degree in diameter. To obtain an undistorted image for this large field, correction-lens systems have to be incorporated and these are located between 20 and 100 cm in front of the photographic plate.

34 The astronomer monitors the operation of the computer-controlled telescope from the control console.

35 The 3.9-metre Anglo-Australian Telescope was inspired by the 3.8-metre telescope of Kitt Peak Observatory in the USA. Both telescopes have a horseshoe mounting. The ring with the prime focus cabin at the front end of the skeleton tube can be completely removed when the optical systems are changed and can be replaced by other units. The control centre can be seen in the background.

36   Mount Stromlo Observatory is part of the Australian National University. The only reminder of its original task, solar research, is a 30-cm telescope but this is no longer in use. The 1.9-metre, 1.3-metre and 76-cm telescopes are employed for stellar observations. A 25-cm photographic zenith telescope is used for the determination of time and for monitoring the rotation of the Earth. A 66-cm Schmidt camera on the site of Mount Stromlo Observatory belongs to the Swedish Observatory of Uppsala. Yale and Columbia Universities also used to operate a 66-cm refrac-tor on Mount Stromlo. Like Siding Spring, Mount Stromlo Observatory was thus also an international institution.

37   Up to 1974, the 1.9-metre telescope of Mount Stromlo Observatory was the largest instrument of the Southern Hemisphere. The observation platform providing access to the Cassegrain focus can be seen below the telescope. The Newtonian focus at the front end of the skeleton tube can be reached via a mobile observation platform which is located in the cut-away part of the dome.

38   The buildings of the Royal Observatory Edinburgh include the main building, erected in Victorian style in 1896, and the extensions of the 1960's. The library wing can be seen extending into the centre of the picture while on the right there is the East Dome with the 90-cm reflector. This was installed in 1932 to determine the energy distribution in stellar spectra by the photographic-spectroscopic technique. This project marked the start of work on astrophysical problems at the Royal Observatory and it is this field of astronomy which is now the exclusive interest of this Institute.

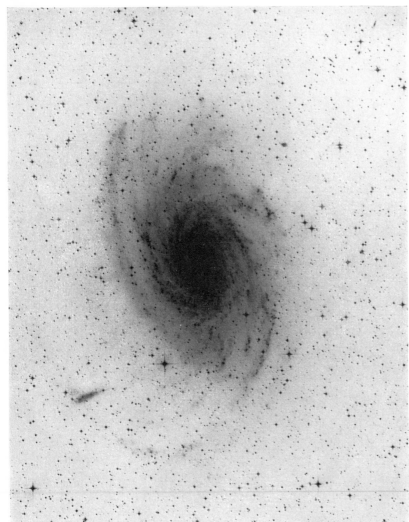

39 a/b   The COSMOS measuring equipment for sky photographs developed by the Royal Observatory Edinburgh in collaboration with industry is exceptionally versatile in its functions and in the data made available. The example shows the quasi three-dimensional map of the blackness distribution on a section of the photographic plate, as produced by the computer. It is the reproduction of a star field photographed with the 1.2-metre Schmidt telescope in Australia. The difference in the image structure resulting from the different brightness of the objects is clearly evident. A striking object is the bright star at the left edge with its diffraction cross produced in the telescope and the halation. This should be compared with the exposure taken by direct photography (not identical with the mapped area).

Page 73:
40   General view of McDonald Observatory on Mount Locke.

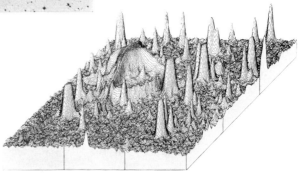

Pages 74/75/76:

41　The 2.7-metre telescope of McDonald Observatory can be used as a Cassegrain and as a coudé system. For the observation of celestial bodies, it is equipped with a coudé spectrograph and several photometers. The 2.7-metre telescope has also been used in a laser experiment for the exact measurement of the distance between the Earth and the Moon. The astro-nauts of Apollo 11 installed a laser reflector 625 sq. cm in area on the Moon. The reflection of the laser beams enables this distance to be determined to an accuracy of a few centimetres. If, in the future, this experiment is carried out from several points on the Earth, it will also be possible to acquire information about continental drifts on the Earth.

42　With the coudé spectrograph of the 2.7-metre telescope and a photoelectric scanner, molecular oxygen was found in the atmosphere of Mars, small amounts of water vapour were detected in the atmospheres of Mars and Venus and the atmosphere of Titan, the largest satellite of Saturn, was discovered.

43　When the 2.1-metre Otto Struve Telescope was commissioned in 1939, it was the second largest in the world, surpassed only by the 2.5-metre Hooker Telescope on Mount Wilson. The 2.1-metre has been modernized in recent years. It was given computer control and equipped for observations in the infrared area. Sensitive detectors were also installed in the spectrographs.

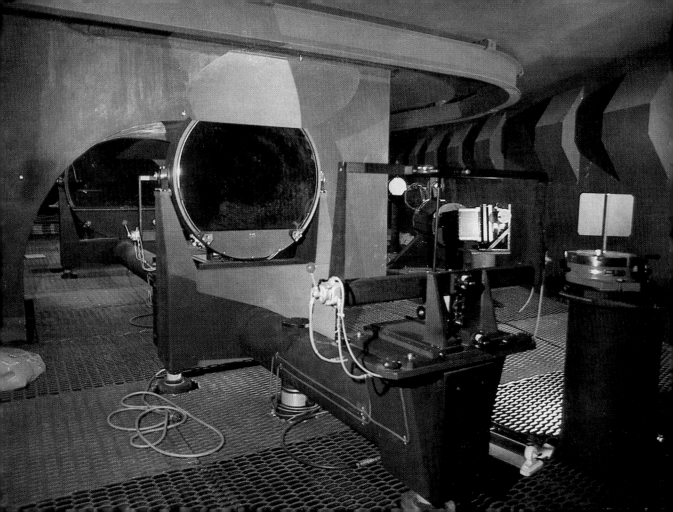

● Within the astronomical sphere, the Sun is a special and relatively independent subject of research. This is due to its closeness—by astronomical standards—which permits studies in great detail, in comparison with stars of low brightness. However, there is no difference in physical terms between the Sun and the stars. The special position of solar research is even evident from the particular construction of the buildings and instruments used. In the course of daylight observations, a significant rise in air temperature can occur. Hot-air distortion then occurs (*schlieren*) which turns the telescope image of the Sun into a vague, quivering disc. One endeavours to counter this problem by a careful selection of the site for an observatory. In addition, the slit of the telescope is placed at as great a height as possible above the surface of the Earth with the aim of avoiding the effect of the disturbances which are restricted to a height of 20 metres above the ground. This is why the term "tower telescope" is typical for solar research. It is precisely at the Kiepenheuer Institute for Solar Physics that great attention is devoted to the question of an objective evaluation of local visibility conditions on the basis of systematic measurements.

The Fraunhofer Institute, now known as the "Kiepenheuer Institute for Solar Physics", was founded at Freiburg im Breisgau in 1943. Its principal task is still the investigation and monitoring of physical happenings in the outer layers of the Sun. About 15 scientists from the Institute concern themselves with questions dealing with the formation of the absorption lines in the spectrum of sunlight, the investigation of sunspots and their magnetic fields, of flow processes in the solar atmosphere and with the oscillations of the entire body of the Sun.

The first instrument used by the Institute—even after 1949 when scientific work recommenced—was the tower telescope on Schauinsland, a 1,240-metre mountain in the Black Forest. However, the climatic conditions soon made it necessary to establish an outstation where there were more hours of sunshine so that the chronological development of activity phenomena on the Sun could be followed more reliably. The choice fell on the island of Capri in the Mediterranean on which the Sun shines for at least four hours daily on more then 200 days of the year. A domeless refractor of the coudé type has been in service there since 1965. With this instrument, a lens of 35 cm diameter always directs the image of the Sun in the direction of the ground—whatever the position of the Sun. This enabled the costly measuring equipment to be installed in a permanent location without having to take account of the changing position of the Sun. To achieve an image of the highest possible quality, the telescope is constructed so that additional protection from the weather is unnecessary. The actual tube mounting with the optical system is enclosed by a separate jacket moved synchronously with the telescope in accordance with the design prepared by the late B. F. Lyot and K. O. Kiepenheuer. Ventilators take care of the circulation of air and thus the cooling in the space between the two assemblies which measures only a few centimetres in extent. From the lens, the light from the Sun passes via two mirrors into the spectrograph which is ten metres in length. With the entrance slit of this spectrograph, it is possible to separate and measure details of only a few seconds of arc in size from an image of the Sun of 35 cm.

Due to the image-forming lens system of the refractor and because of the low sea level, serious light-losses occur during observations in the short-wave spectral region at the Capri outstation so that work in these parts of the spectrum at least has to be restricted. This shortcoming has now been overcome by the establishment of a second outstation by Freiburg Institute. In collaboration with Spanish colleagues, a 40-cm tower telescope was taken into service in 1972 on the island of Tenerife at an altitude of 2,400 metres. This instrument uses reflecting optical surfaces instead of lenses. The optical path within the telescope is in a vacuum—this being provided in the interests of obtaining the highest possible image quality as far as the instrument side is concerned. At first sight, it seems surprising that even in solar research the trend of development is towards telescopes of larger and larger aperture and to more sensitive detection techniques. However, when the solar physicist picks out smaller and smaller details in the disc of the Sun and examines these with a constantly increasing power of spectral resolution, he necessarily comes up against the problem which confronts all observing astronomers: the object studied is not bright enough for the aperture available!

Within the framework of the growing international links in astronomical research, the Kiepenheuer Institute will be playing a major role with the construction of a solar observatory equipped with a 60-cm vacuum telescope on La Palma. This project is being sponsored by twelve European countries on a joint basis.

The step taken by the Freiburg solar physicists from Earth-bound observation to telescopes carried by balloons or satellites was of particular significance for the acquisition of more knowledge about the physical processes on the Sun. These platforms in the sky are not affected by the absorption effect of the Earth's atmosphere and the inconvenience of image instability. These programmes were based on the use of US OSO (Orbiting Solar Observatory) solar satellites and, above all, on the development and testing of the „Spectro-stratoscope" balloon observatory over a period of ten years. This was finally launched in Texas in May 1975 and, in a flight of almost 12 hours at a ceiling of 28 km, sent 400 solar spectra and more than 1,000 photographs of granulation back to Earth. The staff of the Institute will be busy for a fairly long time with evaluating this material.

● In 1932, the American engineer K. G. Jansky discovered that not only optical radiation but also radio-frequency radiation comes from Space and penetrates the atmosphere of the Earth. However, it was only after 1945 that this significant discovery was utilized for astronomical research and that the first great radio telescopes were built.

In 1954, a group of American radio astronomers met at Washington. This meeting submitted a proposal to the National Science Foundation, which had been set up in 1950, that a national radio observatory should be established and that all astronomers should have the opportunity to use it. The task of the National Science Foundation is primarily the support and encouragement of pure research, the provision of assistance for the training of scientists and the exchange and dissemination of scientific information. The National Science Foundation now runs four astronomical institutes: the National Radio Astronomy Observatory, Green Bank, West Virginia, the National Astronomy and Ionosphere Center at Arecibo, Puerto Rico, the Kitt Peak National Observatory near Tucson, Arizona, and the Inter-American Observatory in Chile.

However, before the national institute and in particular the radio observatory could be established, a great deal of work had to be implemented. To preserve the national and open character of the new institute from the very beginning, the National Science Foundation contacted the Associated Universities Inc. for a first feasibility report on the proposal and the utility of a national radio observatory.

One of the important questions which had to be clarified already in the first study was the selection of a suitable site for the National Radio Astronomy Observatory. In optical astronomy, it is desirable to pick up objects of extremely low brightness so that one can see as far as possible into Space. The same applies to radio astronomy. To be able to receive the weak radiation from Space with the minimum disturbance, the area of a radio astronomy observatory must be largely free from terrestrial radio sources. Radio and television stations are real sources of disturbance. Electrical machinery, cars and aircraft can also cause trouble. The site ultimately selected for the National Radio Astronomy Observatory is surrounded by mountains which represent some sort of a screen against interference from terrestrial sources. Furthermore, this is an area with only a small population and little industry. In coordination with the Federal Communication Commission and the Inter-Service Radio Advisory Committee, measures have been taken to keep this area free from radio interference in the future as well. The area selected is in Deer Creek Valley, not far from Green Bank in West Virginia. For the construction of the National Radio Astronomy Observatory, the US Government acquired an area of 11 sq. km.

The basis for the subsequent scientific work was established between 1957 and 1960. This included in particular the construction of the comprehensively equipped electronic laboratories, an essential part of the Radio Astronomy Observatory. The largest building with the administration centre, working facilities for the permanent staff and guests, the electronic laboratories and the library was given the name of the discoverer of radio waves from Space, Karl G. Jansky. The large guest house with a restaurant was built to provide pleasant accommodation and good working conditions for the radio astronomers who come to Green Bank from all parts of the USA and from abroad, too. The recruitment of staff for the Institute was also largely completed between 1957 and 1960. In 1960, the permanent staff comprised 200 scientists, technicians and administrative personnel.

The first large telescopes were designed and built at the same time as the Institute. The first large telescope was taken into service in 1959. It has a diameter of 26 metres and the radiation-collecting surface is completely covered with aluminium foil 3 mm thick. With this very precise reflector surface, observations down to a wavelength of 2 cm are possible. One of the largest

movable parabolic radio telescopes was completed by 1962. It took two years to build and cost one million dollars. It is 92 metres in diameter. Its reflector surface, 7,000 sq. metres in area, consists of a wire mesh. This is why only observations up to a wavelength of 21 cm are possible. Too many of the shorter waves "slip through" the mesh and are thus lost.

Three years later, in 1965, a 42.5-metre radio reflector was taken into service. This is the largest instrument with an equatorial mounting and has a fully covered reflector surface in exactly the same way as the 26-metre telescope. It can be used even for observations down to a wavelength of 2 cm. The continuous reflector surface, the equatorial mounting, and the sophisticated electronic equipment are the reasons why it took seven years to build and cost 14 million dollars, a relatively high sum in comparison with the 92-metre telescope. This instrument was and is preferred for the observation of narrow spectral lines such as those emitted by ionized hydrogen and helium and by the interstellar OH molecule. It has been used for interferometric observations in conjunction with other American and even European radio telescopes. This method of observation (VLBI) is used especially for the determination of very small angular diameters of radio sources.

Of the many other telescopes operated by the National Radio Astronomy Observatory, mention should be made of an 11-metre reflector. It is a small instrument but of the highest precision. This is absolutely essential for observations in the millimetre range, the radio-frequency region with the shortest waves, which is next to the infrared region. Since the telescope was designed for this wavelength area, it was not erected at Green Bank but at the Kitt Peak National Observatory, Arizona, which is situated at a height of 2,000 metres and is the optical counterpart to the National Radio Astronomy Observatory. Whereas radio telescopes are otherwise situated in the open air, this delicate telescope is housed in a dome-like structure. Numerous interstellar molecules have been discovered with this 11-metre telescope, including for example carbon monoxide, silicon monoxide and ethyl alcohol.

If an astronomer wishes to carry out observations at the National Observatory, he must submit his programme to the director, stating the technical equipment required. If the programme has a high scientific content, the visiting astronomer is allocated observing time at a telescope. He may then make use not only of the instrument and its direct ancillary equipment but also of all the technical equipment of the Institute, including the large computer for the reduction of the observational data. No charge is made for the work carried out at the National Radio Astronomy Observatory on behalf of the visiting astronomer.

● To the south-east of Hamburg and on the edge of the suburb of Bergedorf, there lies the observatory of Hamburg University. At a distance of 20 km from the centre of this great city with its mighty port and industrial installations, the Observatory is no longer safe from serious disturbance in respect of its observation work. For this reason, the observational activities of the staff there are being carried out to an increasing extent at institutes with better conditions. This is not the first time that Hamburg's astronomers have moved away from the growing city. In 1833, the Free Hanseatic City of Hamburg took over as a State institute the Observatory which had been founded shortly before at the Millerntor Gate in the port. In 1910, it was greatly enlarged and transferred to Bergedorf. It was made part of the University after the First World War. Its scientific history began with the very practical tasks of a time service, this being associated with the character of Hamburg as an ancient trade centre. In the meantime, the name of the Observatory has become linked with the extensive astrometric work carried out on the *Zonenkataloge* (Position Catalogues) of the Astronom-

ical Society, with the *Bergedorfer Spektraldurchmusterung* (Bergedorf Spectral Catalogue) and, more recently, with research on the structure of our stellar system and on the composition and chronological development of stars with the aid of mathematically formulated model concepts.

The position catalogues referred to here give the exact positions of some 180,000 stars in the Northern Sky. This vast project was planned about a century ago and has been carried out three times since then. The aim of this project was the determination of the slight displacements, the proper motions of the stars as this is known, which can only be obtained by comparisons of the positions determined at different epochs. The project as a whole was of international character but the second and the third survey were based largely on the work and initiative of Hamburg Observatory. A similarly ambitious plan was the photography of the spectra of 150,000 stars in selected areas of the sky and their classification by a spectral type. This 30-years' work was completed in 1953 by the publication of the last of the five volumes in the catalogue. Interestingly enough, the spectral photographs were obtained with the Lippert astrograph, an instrument which an amateur astronomer of Hamburg had really acquired for his private observatory but then, in the interests of more intensive use, had put at the disposal of the University Observatory.

Of the special observation instruments, mention must be made of the Great Refractor with a 60-cm lens having a focal length of 9 metres. Telescopes of this kind formed part of the essential equipment of an efficient observatory at the turn of the century. Its purchase took half of the total budget available for the re-equipment of the Observatory at Bergedorf. Today, the instrument is used for taking astrographic exposures for the optical identification of radio sources. An even more important part of a "classical" observatory was the meridian circle for the determination of the exact positions of stars on the celestial sphere. At Hamburg

Observatory, it was used for the measurement of guide stars as reference points for the photographic work associated with the position catalogues mentioned above. In 1967, this meridian circle was sent to Perth in Western Australia for use in an international programme to study the Southern Sky. Up to that time, the Southern Sky had really been neglected in the field of astrometry, too. The idea was that this shortcoming should be rapidly remedied by the despatch of the appropriate instruments. In the case of the Hamburg instrument, this was preceded by the thorough overhaul of the mechanical parts, conversion to photoelectric measurement and read-off systems and, of course, provision for automatic data output suitable for use with computers.

The Schmidt telescope with a mirror diameter of 120 cm and a corrector plate diameter of 80 cm has been of great importance for astrophysics at Hamburg Observatory. On account of its wide field of view, this type of instrument has proved especially suitable for use in large-scale research programmes. These have included a photometric study of galactic star clusters in several wavelength ranges and a spectral survey of the Milky Way for young, hot stars. The latter are especially useful, on account of their great luminosity, for the investigation of the structure of our star system. Unfortunately, the brightness of the sky above this modern city has a particularly devastating effect on such a high-powered instrument as the Schmidt reflector. It is therefore planned that it should be moved to the Spanish observing station of the Max Planck Institute for Astronomy of Heidelberg. However, the proposed new location of the instrument will require a modified mounting to take account of the more southerly latitude. The original mounting has been used for a 1.2-metre reflector whose purchase has been made possible by a private bequest and which is much less affected by troublesome light sources.

Of the personalities associated with Hamburg Observatory, special mention must be made of J. G. Repsold (1751–1830) and B. Schmidt (1879–1935). Both were associated with the technical side of astronomy. J. G. Repsold founded a mechanical workshop which was subsequently known by the name of "A. Repsold & Söhne" and for three generations represented progress in the construction of astronomical and geodetic instruments. With his meridian circle observations, he was one of the pioneers of serious astronomy in Hamburg and it was also his efforts which led to the establishment of Hamburg Observatory. Of the initial equipment at the Observatory supplied by Repsold's firm, the Great Refractor and the meridian circle are still in use today.

Bernhard Schmidt, who was born in Estonia, worked at Hamburg Observatory from 1926 onwards as a self-employed worker. He was a gifted optical instrument-maker but an individualist who refused to have any firm connection with an enterprise or institution. In 1930, he succeeded in constructing a mirror system which was free from aberrations and with a field of view equalled in size up to that time only by refractors. Intuition and fine craftsmanship were the basis of the idea and execution of the optical system which now bears his name. Numerous telescopes with Schmidt optical systems are now in existence and major progress in optical astronomy of recent decades has been associated with this type of instrument.

● The layman often associates the word astronomy with romantic night-time observations with the telescope and with exciting glimpses of far distant worlds. This is only partly correct. As is the case with every other branch of science, astronomy, too, is obliged to collect numerous facts before it achieves success, in this case a scientific result. The importance of extensive numerical calculations in this is indicated by a glance at the tasks of the Astronomical Calculating Institute in Heidelberg.

The unique character of this establishment is already evident from the circumstances of its foundation. In connection with the calendar reform carried out in 1700 in the Protestant part of Germany standard calendars had to be calculated. These calendar problems led to the founding of the "Societät der Wissenschaften" (Society of Sciences), the later Royal Prussian Academy of Sciences at Berlin, and a calculating institute. In the Appointment Document signed by Elector Frederick III for the first astronomer of the Society, Gottfried Kirch, it states that he should "... fleissig observiren, ... jährliche Ephemerides motuum Caelestium sowohl planetarum alss auch, wenn er darzu Zeit übrig hat, Satellitum calculieren, darneben auch die vermöge Unseres Edicts vom 10. Maji dieses Jahres in Unseren Landen forthin allein gültige Kalender ... jährlich um die richtige Zeit verfertigen und einrichten ..." ("... diligently observe, ... calculate annual Ephemerides of the movements of the planets ... and when he has time, of the Satellites and, according to our Edict of the 10th of May of this year, produce the official calendar every year at the appropriate time ..."). This was the beginning of the Astronomical Calculating Institute which was ultimately located in the Dahlem district of Berlin until the confusion of the Second World War necessitated its complete reconstruction in Heidelberg.

The astronomical calendar (or ephemeris) of these early days became the famous *Berliner Astronomisches Jahrbuch*, which was the astronomers' handbook from 1776 until 1959 and supplied much of the data they needed for their work concerning the Sun, the Moon, the planets and their satellites, the fixed stars, eclipses and occultations of the stars by the Moon. A decision of the International Astronomical Union, responsible for the worldwide coordination of astronomical research, assigned the calculation and publication of such ephemerides to similar institutes in various countries. This superseded the work on the *Berliner Astronomisches Jahrbuch*. However, the Astronomical Calculating Institute was then given the task of the annual preparation of the catalogue on the *Apparent Positions of the Fundamental Stars* since the modern publications no longer contained any data on the fixed stars. This task, too, was in keeping with the scientific traditions of the Institute. Since the time of the First World War efforts had been made there to create a cohesive system of the exact positions of stars to locate the network of astronomical coordinates on the sphere. These fundamental stars as they are known play the same role in the sky as the trigonometrical points on Earth but the problem is very much more complicated on account of the proper motions of the stars and the complex movement of the Earth. In the course of time improved reference catalogues have been produced on the basis of standard measurement techniques for the determination of positions by other observatories. Another fundamental catalogue known as the *FK 5* is already at the planning stage. Without these efforts in the field of positional astronomy which might appear somewhat dry to the outsider, essential and perhaps apparently more interesting astronomical and astrophysical work would be condemned from the very start.

Tasks such as the annual publication of almanacs or ephemerides and the calculation of the orbits of small planets could be done as routine work and thus involve the risk of stagnation. This was the reason, even at that time, why staff were encouraged to concern themselves with topical problems. This explains why the scientists at the Astronomical Calculating Institute, now 16 in number, have also published articles over a period of many years which examine the movements in the planetary system and in stellar systems from the viewpoint of the driving forces and are not solely concerned with a kinematic description. Questions of the structure and development of stellar associations, such as that in the vicinity of the Sun, of star clusters and galaxies, are being successfully investigated.

Finally, mention must also be made of a major contribution by the Astronomical Calculating Institute to astronomical research: the detailed compilation of the international work of reference *Astronomy and Astrophysics Abstracts*. Since 1969 this publication, which was previously known for seventy years as the *Astronomischer Jahresbericht*, provides astronomers at six-monthly intervals with a classified review of everything published on the subject in the course of the half-year in question. This is of invaluable assistance for tracing works of specific interest among the vast number of astronomical papers now published.

Computer technology naturally plays an important part in all the activities of the Astronomical Calculating Institute. In particular, it is the high speed of the computer which permits the solution with a high degree of accuracy of the mathematical equations associated with scientific problems. This is of special significance for the handling of dynamical problems. On the other hand, the great memory capacity of computers is utilized for the processing of the extensive observational material evaluated in catalogue work. All the available data is stored and is automatically called in the course of processing. Largely computer-controlled techniques are also used for the preparation of the bibliography volumes of the *Abstracts* which are published twice a year. Punched card machines were formerly used but the first electronic computer in Heidelberg was taken into service in 1961 at the Astronomical Calculating Institute. In the meantime, the Institute has an even larger machine which is also linked with the large main-frame computer of Heidelberg University and can also use the capacity of the latter.

● On the wooded slopes of the Königsstuhl, at an elevation of 564 metres near Heidelberg, there are two independent research institutes: the Regional Observatory, which was built between 1896 and 1900, and the Max Planck Institute for Astronomy which was previously part of the former but which was organized as an independent institution in 1969. Although they have separate budgets and staff, sensible and task-related links are still maintained. Their relative proximity and the origin of the Max Planck Institute are contributory factors in this.

The establishment of an astronomical research institute within the framework of the Max Planck Society was carried out with the aim of achieving a substantial improvement in the position of optical astronomy in the Federal Republic of Germany. The plan called for the establishment of an efficient administrative centre in Heidelberg, and the setting up of an outstation with several large telescopes in the Northern Hemisphere and an outstation with a 2.2-metre telescope in the Southern Hemisphere. The first two parts of the Institute have been completed but the project for a southern observatory has had to be postponed. For the observatory in the Northern Hemisphere a site had to be chosen which was within easy reach but nevertheless offered outstanding conditions for observation so that the high investments associated with it were justified. Consequently, for a period of several years, possible sites in Greece and Southern Spain were carefully and objectively compared. The location finally chosen was the mountain of Calar Alto in the Spanish province of Almeria to the east of Granada. The construction of the observatory there was laid down within the framework of an agreement at government level by which Spain provided 100 sq. kilometres of terrain and undertook to develop this area.

The Max Planck Society, on the other hand, was responsible for the erection and operation of the buildings with the scientific installations. Spanish astronomers also take part in the research work and at least

10 % of the observing time at the telescopes is at their disposal. In addition to the construction of the necessary accommodation and supply facilities, a 1.2-metre telescope and the 1.5-metre telescope of Madrid Observatory were taken into service by 1978. At that time, the work on a 2.2-metre telescope had not been finished. The dome building with its steel dome 20 metres in diameter was finally completed after a delay, while the actual telescope had already been handed over to the Max Planck Institute and was awaiting erection. The construction of the building for the reflecting telescope with an aperture of 3.5 metres as planned was started at the same time but some years will pass before this instrument, too, can be taken into service. Eventually, the equipment will also include a German guest: the large Schmidt reflector of Hamburg Observatory will be moved from Hamburg-Bergedorf to Southern Spain since at its present site there are no longer the proper observation conditions for this high-power instrument.

The 1.2-metre telescope is used largely for spectroscopic and photometric work. Its optical design is of the Ritchey-Chrétien type and it has a field of view almost 1° in diameter. The photometric equipment is mounted at the Cassegrain focus behind the pierced main mirror, the spectrograph is permanently located on one of two arms of the fork mounting and the rays of light from the celestial body viewed along the optical axis of the telescope are reflected if required through the fork and into the spectrograph. In the case of the 2.2-metre telescope, there will be not only a Cassegrain focus but also a coudé arrangement, which, through skilful control of the light-path, requires only four mirrors to direct the light into a vertical spectrograph underneath the dome room.

The Max Planck Institute for Astronomy has invested a great deal of money in ancillary telescope equipment and in the radiation detectors. For example, modern electronic image-detecting systems are used for effective use of the incoming rays of light. Photometry for wavelengths in the infrared range has already been firmly established at the Regional Observatory. This is why the Observatory has developed for Mount Calar Alto computer-controlled measuring equipment, which operates in the infrared range as far as this is possible under atmospheric conditions, for determining intensity and polarization.

Of the work carried out on infrared astronomy of a wide variety of objects, special mention must be made of the measurements of dark clouds and of areas of ionized hydrogen. This is associated with the investigation of extremely red objects which are physically associated with proto-stars. At wavelengths greater than approximately 12 μm, the molecular gases present in the terrestrial atmosphere complicate measurements made on Earth. Hence balloon experiments have been carried out at Heidelberg for some years now, data of the infrared radiation of the Galaxy, approximately at wavelengths of 50 μm, being obtained at an altitude of 35 km. These experiments are also used for the investigation of zodiacal light at selected shorter wavelengths and in the ultraviolet spectral region. Zodiacal light is a low-intensity effect of light which occurs through the scattering of sunlight on dust particles in interplanetary Space. Heidelberg has always made important contributions to the study of this form of cosmic matter. Since the end of 1974, the Heidelberg astronomers had had entirely new opportunities in this respect, when the space-probe Helios 1 was launched in the USA for carrying out experiments in the area around the Sun. Since this spacecraft attained the minimum distance from the Sun technically possible at the present time (the point nearest the Sun is within the orbit of Mercury), the Heidelberg astronomers obtained important new data on the zodiacal light as well.

44 This special instrument is in operation on the island of Capri. The outstation of the Kiepenheuer Institute for Solar Physics of Freiburg is situated in classical surroundings, 150 metres above the celebrated Blue Grotto and not far from the ruins of the palace of the Emperor Tiberius. The special design of the instrument removes the need for a protective dome and thus avoids image distortion from air turbulence at the slit. The same considerations have led to the location of the telescope on a 11 metre high building which simultaneously acts as an instrument support.

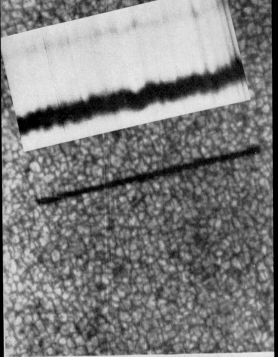

FRAUNHOFER INSTITUT FREIBURG   PROJEKT SPEKTRO-STRATOSKOP
Flug am 17. Mai 1975 in Palestine / Texas. Bild von der Granulation
in der Mitte der Sonnenscheibe im weißen Licht (λ 5560 Å) mit Spalt
und Ausschnitt aus dem Spektrum bei λ 5130 Å. Die Länge des Spaltes
beträgt 60".

45   In the observation room
of the Schauinsland Solar Observatory the image of the Sun is
formed on the slit of the spectrograph, it then passes through
an enclosed grating spectrograph of eight metres in length
and is finally available in this
room as a spectrum of high resolution for various kinds of
measurements. The direct image
of the Sun can be taken from
the slit by a TV camera (centre
of picture) before it reaches
the spectrograph and transmitted to a monitor or tape. Two
pairs of photocells control the
position of the image of the Sun
on the slit with the aid of control mechanisms and thus ensure
that a detail under observation
retains its exact position for the
whole of the period of measurement.

46   The surface of the Sun
reveals closely packed high-
temperature mass-elements rising up side-by-side with cooler
elements which are sinking in
the opposite direction. This
quickly changing finely detailed
structure, called granulation,
is known from terrestrial observations.

47 At the end of the 1930's, G. Reber, a young engineer and amateur radio operator, set out to check and supplement the results obtained by G. Jansky. For this, he built a radio telescope of 10 metres in diameter behind his house at Wheaton, Illinois. Since his investigations with this telescope were of great importance for radio astronomy, the instrument was relocated under his supervision on the terrain of Green Bank Observatory. The Grote Reber is still used today at Green Bank for the testing of new detectors.

48 The first test observations with the 92-metre telescope were made on 21 September 1962. 600 tons of steel were used for this telescope and its total moving mass amounts to 500 tons. It can be moved at a maximum speed of 10° per minute.

Hamburg Observatory
Federal Republic of Germany

49   This aerial photograph of Hamburg Observatory shows the 14-metre dome of the Great Refractor in the foreground. On the right behind it, there is the building of the Schmidt reflector opened in 1959. This instrument has since been relocated in Southern Spain and its place taken by the Oskar Lühning Reflector. At the right edge of the picture, the barrel-shaped roof of the meridian circle will be seen, the design of which results from the directional nature of the instrument. On the outer left, there are the principal building and the small houses for the staff of the Observatory.

50   The first Schmidt camera in the world was constructed at Hamburg in 1930. Its reflector is 44 cm in diameter, it has a focal length of 62.5 cm and the diameter of the corrector plate is 36 cm. This reflector was the first instrument to combine high speed with a very large field of view (15°). Unfortunately, the original mirror was destroyed during the war and had to be replaced but the corrector plate and the tube are from the original instrument.

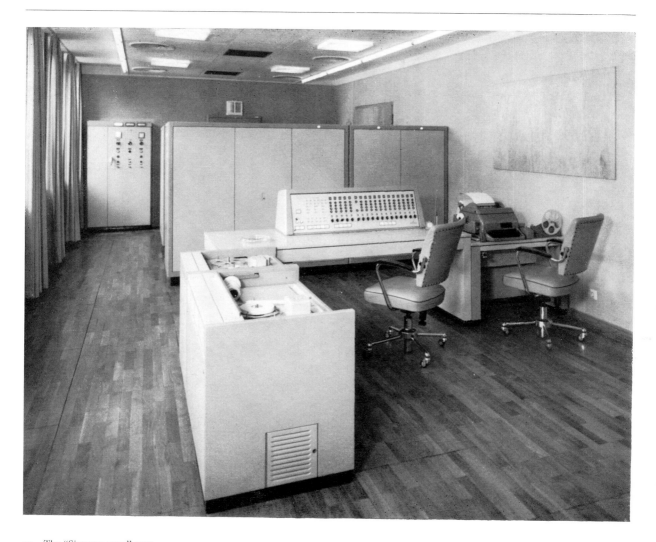

51 The "Siemens 2002" computer installation was used for
over 15 years for scientific tasks
and the catalogue work of
Heidelberg Astronomical Calculating Institute. In the centre
of the picture, there is the control console, flanked on either
side by periphery units, while
in the background there are the
electronic cabinets, still of a
large size at that time.

52 The process of separating
the Max Planck Institute for
Astronomy from the Regional
Observatory was completed
with the dedication of the new
institute building on the Königsstuhl near Heidelberg in 1976.
At the front right of the picture,
there is the Astro Laboratory
with two domes for instruments
with an aperture of about 70 cm
which can be used for the testing of auxiliary instruments in-

tended for Calar Alto. The
buildings of the Regional Observatory can be seen behind
the strip of woodland.

Page 91:
53 Baden Regional Obser-
vatory on the Königsstuhl
near Heidelberg was founded
by M. Wolf (1863–1932). For
astronomy, his name is associated with pioneer work in astrophotography and with investi-

gations of basic importance
using the photographic method,
especially in the area of our
stellar system.

54 This picture shows the
Helios 1 probe shortly before
completion in the laboratory.
Solar cells covering the outer
surface of the double cone supply energy. Various antennas
extend upwards for communication with ground-control sta-

tions. The zodiacal light photometer developed by the Heidelberg research group cannot be
seen since it is installed in the
large opening of the cone on the
underside in the picture.

Page 92:

55   The photograph shows the
gondola of the Thisbe balloon
telescope operated by Heidelberg astronomers, prior to
launching at the balloon launch-
ing-site at Palestine, Texas.

56   Buildings of the Max
Planck Institute, Heidelberg, in
the Spanish province of Almeria. On the left, there is the
2.2-metre building, in the foreground the 1.23-metre building
and in the background the building of the 1.5-metre telescope
with its square ground-plan.

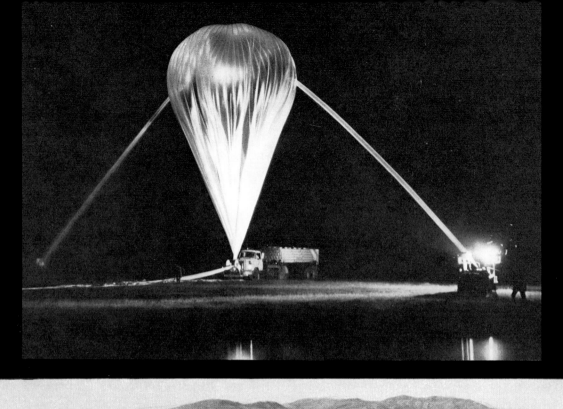

● Without doubt Greenwich Observatory is one of the most famous observatories in the world, even for laymen. This is because of its very long history, its importance for navigation during the time of English maritime power and, in particular, because the Zero Meridian and the reckoning of Universal Time were once fixed on the basis of the position of the Observatory. Today, the public interest that it attracts is more general on account of the time service provided by the Royal Greenwich Observatory, the publication of yearbooks such as the *Nautical Almanac* with their importance for the preparation of calendars, surveying and navigation and through the monitoring of the Sun which is carried out in the interests of radio traffic. A total of some 230 staff here are making an acknowledged scientific contribution to classical astronomy, astrophysics and the development of astronomical equipment. The main building of the original observatory, which is now used as a museum, in the London suburb of Greenwich, bears a Latin inscription stating that King Charles II of England, the Patron of astronomy and navigation, founded this observatory in 1675 to promote the sciences. The development of navigation was hindered in particular by the difficulties encountered in determining the position of a ship. It is true that the latitude could be determined from the elevation of the North Star but the exact positions of stars and a zero point for the computation of coordinates were lacking for the determination of longitude. This was one of the tasks expressly laid down at the time the Observatory was founded. The site selected was on a hill in the Royal Park of Greenwich, close to the River Thames. The development of the buildings and instruments and the history of its scientific achievements are associated with the names of great astronomers. J. Flamsteed was the first to bear the title of Astronomer Royal. He was followed by E. Halley, familiar to us from the determination of the orbit of the comet named after him, and by J. Bradley as the actual founder of the scientific reputation of the Observatory. His

work met with such approval that the royal administration of the Observatory granted his request for an assistant.

The instruments at Greenwich Observatory were arranged for the determination of the positions of stars. Although there were refractors there with a long focal length, the instruments typical of it were the mural quadrant and the astronomical transit circle. Instruments of this type can only move in the meridian plane and they supply the coordinates of a star from a determination of the exact time and elevation of its meridian passage.

After the Second World War, the Observatory had to be moved on account of the marked deterioration in observation conditions resulting from urban development. The historical buildings were taken over by the National Maritime Museum. A part of this is open to the public at the present time but most of the instruments there are only replicas. However, numerous visitors are especially attracted to an original instrument, the meridian circle of Sir G. B. Airy, the seventh Astronomer Royal. Since 1851, this instrument has been used for more than a century for the measurement of times and elevations of passage. More than 600,000 measurements have been made with it. Its attraction for visitors is due to the fact that this instrument defines the Prime (Zero) Meridian. It is here that the Earth is divided into the Eastern and Western Hemispheres, as emphasized by a copper strip laid in front of the building. To be sure, the location of this meridian is only by virtue of agreement and is not designated by astronomical features. After a number of different definitions had been used, the International Meridian Conference in Washington agreed in 1884 to calculate longitude on the basis of Airy's meridian circle. This finally legalized the reference line which had already been chosen at that time for more than 70 per cent of all marine charts.

Since 1948, the Royal Greenwich Observatory has been housed in the castle of Herstmonceux in southwest Sussex, a few kilometres away from the Channel coast of England. This move was rewarded with very much better observation conditions and, since the University of Sussex is not far away, contact with other scientific institutions is ensured, a necessary condition for modern research. Administratively, the Royal Greenwich Observatory now comes under the aegis of the Science Research Council, a body for the promotion of scientific work in the United Kingdom. It is therefore understandable that the Observatory is now charged with a more comprehensive task, being responsible for the provision of opportunities for observation and research for British astronomy in general. This is why some of the smaller telescopes are operated as national instruments, especially the Isaac Newton Telescope which was taken into service in 1967. The observing time on this instrument is allocated by a special committee on the basis of the importance of the programme in question. An outstation of the Observatory on La Palma, one of the Canary Islands, has a similar status. In 1979, the Isaac Newton Telescope of Herstmonceux was transferred there so that better observation conditions there can be used for this instrument, which was also provided with a new Cervit mirror.

The ultimate aim is the construction of a giant telescope with an aperture of 4.2 metres, the design work for which is now being carried out. The intensive preparatory work involved in this includes not only the telescopes but also the instrumentation. After all, the technical effort expended on gathering the rays of light at the telescope is only worthwhile when the latter is used in conjunction with sophisticated detector equipment. Working-groups have therefore been formed which are specifically concerned with the development and use of light detectors, electronic ancillary equipment, optical ancillary equipment and computers. The technical results of their work are reflected in the actual astronomical research projects which at this Observatory are mainly associated with direct observation. Photometric and spectroscopic measurements predominate in this but cover almost all kinds of astronomical objects. Thus the investigations are concerned with the most remote galaxies just as much as with the stars in the immediate vicinity of the Sun, for example. In the case of the galaxies, mention must be made in particular of the photometric monitoring of the irregular changes in brightness which occur with some types and the difficult identification of sources of radiation, initially discovered by radio telescopes, with their optically visible counterparts. Of course, in the area of extragalactical research, the detailed study of individual objects such as the Magellanic Clouds with instruments in the Southern Hemisphere also plays a role. As regards the objects of research within our star system, an interest is taken, for instance, in the globular star clusters, interstellar matter and naturally in the individual stars. The latter are examined from two aspects. One direction of study considers each star as an astrophysical object as such in which the chemical composition of the outer layers, the structure and the evolutionary development has to be investigated. A notable development is the method employed to compare the actual spectra observed with spectra which have been established from theoretical models and in this manner achieve an improvement in theoretical concepts. The other aspect is of a more collective or statistical nature. In this, the stars are regarded only as "particles" which, by virtue of their spatial distribution and the components of their movement, comprise the structure of our stellar system. The interest of the astronomers at the Royal Greenwich Observatory is concentrated in particular on the behaviour in our more immediate cosmic environment.

● The terrestrial atmosphere has a detrimental effect on many astronomical observations. It is consequently understandable that when new observatories are built today, efforts are made to find a site for them at a great height above sea level. An outstanding example in this connection, in the truest sense of the word, is an outstation of the Institute for Astronomy of the University of Hawaii. It is located at a height of 4,205 metres on the peak of Mauna Kea, a volcano on the island of Hawaii which erupted for the last time some 4,500 years ago. At this altitude, 40 per cent of the atmospheric air is below the observatory and the astronomers here therefore enjoy outstanding observation conditions. This begins with the large number of clear nights. This is because the bare, stony peak rapidly cools down after sunset. The downward winds which then occur bring cold dry air from the higher layers and thus quickly break up a slight cloud ceiling which may be present. The atmospherically determined quality of the star images is also of importance for astronomical work. Small, sharply defined star discs in the telescope mean a high concentration of light and thus great range for every kind of measurement. Under such conditions, even delicate structures in extended astronomical bodies can be identified. This is demonstrated by the excellent photographs of planets which have been taken in large numbers from the peak. An especially favourable feature of the site is the low level of water vapour in the atmosphere. Water vapour generally has a very negative effect on the wavelengths within reach on the red part of the spectrum. The infrared radiation, which is of exceptional interest for many investigations, can therefore not normally reach the telescope since it is already absorbed in the higher layers. Measurements taken from Mauna Kea, however, have shown that the level of water vapour here is only 10 per cent of that at sea level. This explains the increasing international interest in the erection of infrared telescopes at these heights and the fact that one third of the observation time on the 2.24-metre telescope of the Institute

was allocated to this spectral area within a short time of the instrument being taken into service. After all, the infrared investigations carried out by staff at the Institute and visiting astronomers bring information about the properties and temperatures on the surfaces of even remote planets and their satellites, they extend our knowledge of the cool envelopes of dust which surround certain types of stars and objects in the pre-stellar stage of development and they penetrate through thick clouds of dust to the centre of our Galaxy which, in the range of visible light, is concealed from even the largest telescopes.

Measurements of the daytime brightness of the sky which were carried out in the 1950's indicated that there were excellent conditions for astronomical observations on Mount Haleakala, a volcano on the Hawaii island of Maui. The dryness and purity of the atmosphere cause a relatively slight scattering of light and make this site particularly suitable for researching light phenomena of low intensity, such as those occurring in the corona of the Sun, the zodiacal light and the light of the night sky. As a result, Mees Observatory was built in 1963 at a height of 3,000 metres and, with the appropriate instruments, specializes in the investigation of these manifestations. In the meantime, the Lunar Ranging Observatory has been built on Haleakala. At this Institute, laser rays are used for determining the distance of the Moon to a very high degree of accuracy, thus providing the foundation for an improved theory of the motion of the Moon. The principal item in the measuring equipment there is a laser which emits three light pulses every second with a high directional concentration and of extremely short duration. These pulses are emitted via an optical system in the direction of the Moon from where they are reflected back to Earth by a special mirror and received again at the Observatory. A time-measuring system then determines the time taken by the light to an accuracy of a ten-billionth of a second and thus establishes the distance of the Moon to an accuracy of some metres. The mirror was placed on the

Moon by Apollo astronauts during the US programme of landings on the Moon.

The Observatory on Mauna Kea was built as another outstation of the Institute for Astronomy of the University of Hawaii at Honolulu after the American space authority NASA had shown its interest in the erection of a giant telescope in connection with its planetary programmes. For this purpose, the late G. P. Kuiper (1905–1973), a planetary expert, also investigated possible sites on the island of Hawaii and discovered the peak region of Mauna Kea. After NASA had provided resources in 1965 for the erection and operation of a 2.24-metre telescope for researching the planetary system, the University of Hawaii built the research station on the mountain which, to begin with, began work with two 60-cm telescopes in 1968. The giant telescope with two optical systems was taken into service in 1970.

9 This cross-section shows the dome building of the 2.24-metre telescope at the Astronomical Institute of the University of Hawaii on Mauna Kea. In the upper storey of the annex, there is the coudé room for the bulky spectrograph installation which is linked with the light-path of the telescope via the polar axis and reflecting mirrors. On the floor below, there is the aluminium evaporation plant used for applying a new reflective coating to the mirror when necessary.

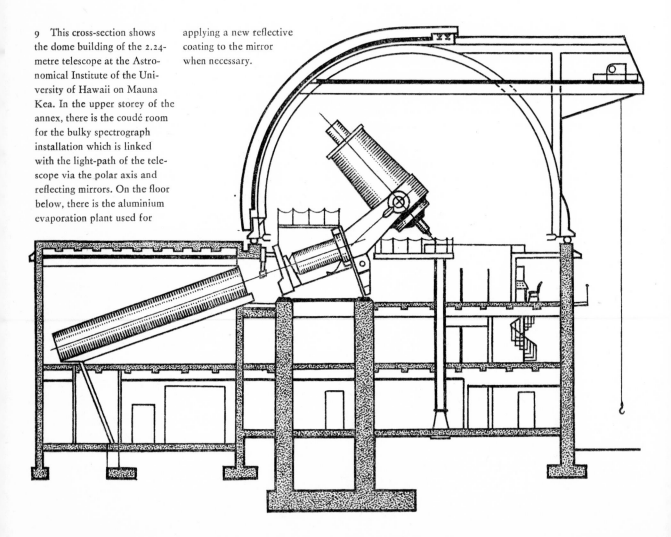

The ideal conditions on Mauna Kea on the one hand and the increasing disturbance of established centres of astronomical research by urban developments, industrialization and air pollution on the other explain why the Mauna Kea region is becoming a stronghold of observational astronomy. Three instruments of the 3-metre to 4-metre class are to be built there or have already been completed. The interest taken by French astronomers in this site led to a programme on a cooperative basis involving Canada, France and Hawaii with a 3.6-metre reflector. The United Kingdom has built the largest telescope (UKIRT) especially designed for observations in the infrared range having a mirror with a 3.8-metre diameter and, finally, NASA has financed another infrared telescope with a mirror diameter of 3 metres.

Living at a height of more than 4,000 metres above sea level is a severe strain for the persons engaged in astronomical work. It is true that the winter on the peak is not characterized by extreme cold but the snow and icy storms make it fairly severe. After the access to the mountain, which is difficult enough under normal conditions, was blocked from time to time by masses of snow during the first winters, heavy snow-clearing machines are now used so that no night suitable for observation is lost. The lack of oxygen at this height means that both physical and mental efficiency is impaired. Before starting work, observers can therefore acclimatize themselves in the "Hale Pohaku", the "stone house" with rest and accommodation facilities, which is situated 1,200 metres from the peak.

● There are far fewer radio observatories than optical observatories. Furthermore, the radio observatories are very recent in comparison with most optical observatories since it was only in 1932 that radio signals from Space were discovered and for the first decade little attention was paid to them. Jodrell Bank Radio Observatory, which was founded in 1945 by the University of Manchester, is consequently one of the oldest establishments of this kind in the world. The Institute is situated 30 km from Manchester. Knutsford is 10 km away and Macclesfield 16 km away. The fully steerable, parabolic radio telescope, for many years the largest of its kind, was taken into service in 1957 at Jodrell Bank. Its radio reflector is 76 metres in diameter and has a focal length of 22 metres. Although the reflector diameter is of such a great size, even the smallest optical telescope has a resolving power superior to this giant radio telescope. This is because the resolving power decreases with the ratio of wavelength to reflector diameter. However, the wavelength of radio waves is about ten million times greater than that of optical radiation. Consequently, a radio telescope would have to have a diameter ten million times greater than that of an optical telescope to achieve the same resolving power. The 76-metre telescope cost £ 700,000 and it was modernized in 1971 at the cost of a further £ 600,000. From 1957 to 1971, it was known as the Mark I telescope and since 1971 as the Mark IA.

The telescope has an altazimuth mounting and the axis of rotation for the elevation adjustment is at 50 metres above the ground.

Two 36-kW motors each are provided for the elevation adjustment and the azimuth motion. Maximum speed is approximately 15 degrees per minute which is high when it is considered that the total weight involved is 3,200 tons. The telescope can operate on wavelengths of 10 cm upwards.

The Mark II telescope was commissioned in 1964. Its radiation-collecting surface is likewise a paraboloid but it is elliptical instead of circular. The larger diam-

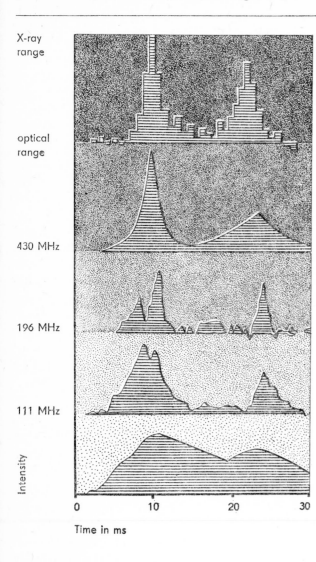

X-ray range

optical range

430 MHz

196 MHz

111 MHz

Intensity

Time in ms

eter is 38 metres, the smaller one 25 metres. This telescope also has an altazimuth mounting.

The Mark III telescope was put into service only two years after Mark II and its dimensions are the same as this instrument. It is situated at Wardle near Nantwich, 24 km to the south-west of Jodrell Bank. The motion of the instrument is controlled and monitored from Jodrell Bank via a radio link.

Another parabolic radio reflector with an altazimuth mounting and a diameter of 25 metres was erected in 1977. This 200-ton instrument is at Knockin near Oswestry, 67 km to the south-west of Jodrell Bank. This telescope is characterized by a very accurate reflecting surface and can operate with a high degree of precision on the 1.2-cm wavelength upwards.

In addition to the above instruments, Jodrell Bank Radio Observatory also has the 25-metre telescope of the Royal Radar Establishment at Defford, 127 km to the south of Jodrell Bank, at its disposal. All the telescopes, including the reflector at Defford, can be controlled by computer from Jodrell Bank via a radio link.

A single radio telescope has a low angular resolving power but if two telescopes are used for observations these two instruments can be regarded as two points of a very large telescope and, in the calculation of the resolving power, the distance between the two telescopes is taken as the diameter of the "giant reflector". If two telescopes are 100 km apart and work on the 5-cm wavelength, they have a resolving power equal to that of a 1-metre optical telescope. This circumstance is exploited at Jodrell Bank and all the telescopes mentioned are jointly operated as a multi-telescope radio interferometer.

With this array arrangement, the angular structures of distant galaxies are investigated with a resolving power of 0.03 to 3 seconds of arc. In radio astronomical observations, the positions of radio sources can be determined to an accuracy of 0.01 seconds of arc. This is also very important for the optical identification of radio sources. Jodrell Bank Radio Observatory has

10   The Crab nebula came into existence as the remnants of a supernova outburst which, according to historical oriental sources, was observed on 4 July 1054. In 1928, it was demonstrated that the nebular matter in the gaseous state was expanding in all directions at a speed of 1,000 km per second.

One of the first pulsars was discovered in the Crab nebula. Its special characteristic is that with 30 double-pulses per second it possesses the highest observed frequency.

very wide-ranging work programmes. The distribution of neutral hydrogen in the Milky Way system, in the Magellanic Clouds and in other galaxies is under investigation. The position, structure and signal polarization of radio sources are being observed and attempts made to identify them optically. The pulsars are one of the most spectacular discoveries made by Jodrell Bank. This new class of cosmic bodies discovered by radio astronomers at Cambridge was observed in 1967 when attempts were being made to demonstrate interplanetary scintillation. Pulsars emit sharply defined radio pulses with a high time constancy. In the case of the approximately 300 known pulsars, periods ranging between 0.03 and 4 seconds have been observed. It is now believed that pulsars are rotating neutron stars characterized by densities of 10^{15} g per cm^3 and diameters of 10 to 20 km. Their existence was predicted in 1932 by the Soviet physicist L. Landau.

In addition to the telescopes already mentioned, there are also two 15-metre radio telescopes in use at Jodrell Bank. One of them is employed for maintaining links with lunar and planetary probes while the other is utilized for radar measurements with the Moon.

Another noteworthy fact is that a planetarium has also been opened at Jodrell Bank and that four programmes a day are held there. The Institute is open to the public every day from March to October and many people make use of this facility.

● Many an astronomer, working in poor weather conditions, has certainly wished that he could set up his telescope above the clouds. Consequently, the idea of making astronomical observations from on board an aircraft is not so far-fetched. At an altitude of 12 km, 99 % of the terrestrial atmosphere is below the observer. This provides excellent opportunities for radiation measurements in the infrared spectral region in particular, since they are otherwise subject to the strong absorption resulting from the water vapour in the atmosphere.

Since vibration-free astronomical observation from an aircraft had to be possible in the Space Age, NASA supported a programme with this aim as long ago as 1965. This did indeed enable successful infrared measurements to be made and thus encouraged further developments in this direction. These efforts resulted in the NASA airborne observatory, named after G. P. Kuiper (1905–1973), which became operational in 1975. Its name honours a researcher who made outstanding contributions to various fields of astronomy and took a special interest in our planetary system and infrared astronomy. G. P. Kuiper linked both these fields of activity with each other in an extremely successful manner. He initiated the first NASA programme for photometry at infrared wavelengths from an aircraft-based position and made a major contribution to all US space missions concerned with planetary research.

The heart of the airborne observatory is a 91-cm reflector which is installed in the fuselage just ahead of the wings and is operated from there as if from the open dome of an observatory. To avoid any absorption of the incoming radiation, no optical window is provided on the outer skin of the aircraft. When in use, the instrument is therefore subject to very low air pressure and to temperatures of about −55 °C as found at altitudes of 12 km. Account of these external conditions had to be taken in the construction of the telescope and it was for this reason that materials such as

Cervit and Invar were selected which vary but little when exposed to greatly fluctuating temperatures. In its optical design, this telescope is of the Cassegrain type with the beam of light bent by an auxiliary mirror and with the focus located in the cabin of the aircraft. It is there that the instruments are operated under almost normal air-pressure conditions, most of them being photometers which detect and process the incoming radiation. So that the noise radiation from the celestial background can be subtracted from the total signal, one of the secondary mirrors is kept oscillating when observations are in progress. Through this, the field of view varies with the corresponding frequency from the object under observation to the background, enabling the signal to be corrected. This is the usual procedure with infrared telescopes.

Despite the great smoothness in flight of a large jet aircraft, special attention had to be paid in the design to directional stability and freedom from vibration in the aircraft/telescope system. This begins with the automatic control of the aircraft. This not only reduces fluctuations but can even be fed with correction signals to keep the plane on a long, curving course so that the object studied can be kept in the field of view despite the apparent daily movement of the sky.

For the active control of the telescope in the event of directional deviations of any kind, an autoguider system operating via an auxiliary telescope is provided which is orientated on the position of two stars on a TV monitor and can thus compensate not only for shifts but also for rotational movements of the system. Even in unsteady flight conditions, directional errors are kept to less than two seconds of arc as a consequence. Finally, the telescope is stabilized against vibrations by pneumatic shock absorbers and an air bearing.

An aircraft is an excellent base for observing eclipses. For Earth-bound observers, an eclipse of the Sun is limited to a bare eight minutes at most but an aircraft can follow the umbra as it moves across the surface of the Earth, thus permitting more protracted pro-

grammes. Notable observations have been made from the G. P. Kuiper Observatory concerning the occultation of stars by planets and asteroids, allowing conclusions to be drawn as to the diameter of the planet in question and the properties of its atmosphere. An outstanding scientific result in this connection was the discovery that the planet Uranus is also encircled by a ring system. This was found during the observation of a stellar occultation in March 1977 when observers in the flying observatory noted periodic diminutions in light from the eclipsed star before and after the actual occultation by the disc of the planet.

The flying observatory is used for about 80 research flights per year, each flight lasting from seven to eight hours. The observation time available with this facility, which is administered by the Ames Research Center of NASA in California, is also at the disposal of scientists from other American and foreign groups. Thanks to the versatility of the telescope, the programmes carried out so far have been very wide in scope. They mainly concerned the photometric and spectrophotometric examination of all types of cosmic infrared sources from the planets to extragalactic stellar systems but have even extended to the investigation of the spectra of interstellar molecules on wavelengths in the millimetre range.

57 The historic buildings of Greenwich Observatory—here Flamsteed House—are now an attraction for numerous visitors. On the little tower, there is the time-ball, the fall of which marked the hour of noon for boats on the nearby River Thames.

Pages 102/103:

58 This is an aerial view of Herstmonceux Castle in Southern England. It is now the home of the Royal Greenwich Observatory. Dome buildings are apparent in the background, the one standing by itself on the right being the 30 metre high dome tower of the Isaac Newton Telescope.

The principal mirror of the Isaac Newton Telescope with its aperture of 2.5 metres was fabricated some years ago from Pyrex, a material with a low coefficient of thermal expansion, in the USA as a test casting for the mirror of the 5-metre telescope of the Hale Observatories in California. A Cervit mirror will be used when the telescope is re-built at La Palma, Canary Islands.

59 The photograph shows an observer behind the pierced principal mirror of the Isaac Newton Telescope at the Cassegrain focus.

60 Greenwich time signals are sent out from this control console. The time service is based on the Observatory's own observations, atomic clocks and comparisons with international time signals.

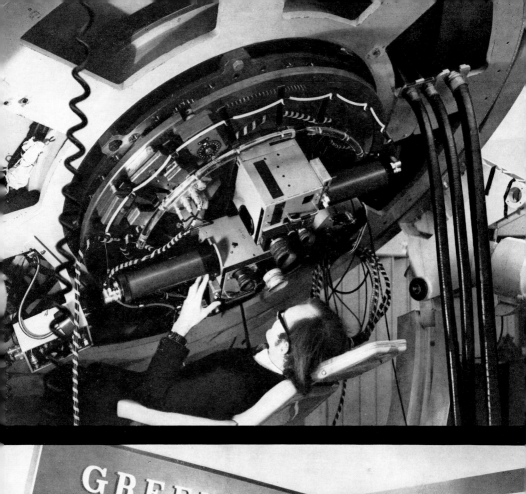

61 The rot
relative to t
of our time
the photogra
scope of the
Observatory

62 An astronomer at work
with the Cassegrain spectrograph of the 2.24-metre telescope of Hawaii University's
Astronomical Institute on
Mauna Kea.

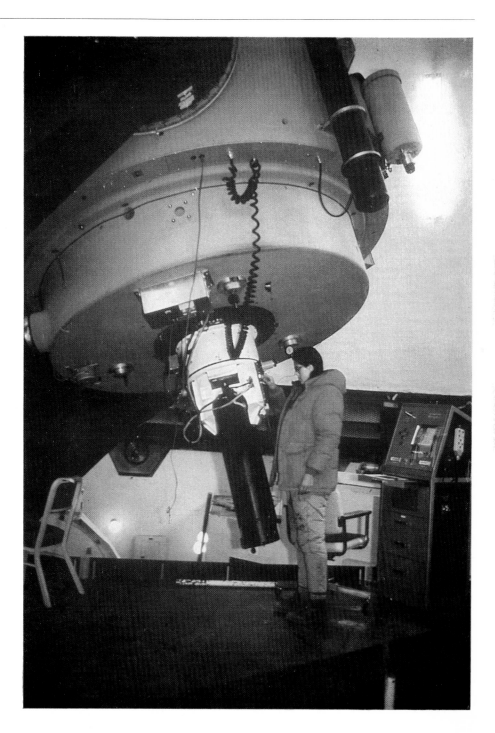

63   The 76-metre radio telescope at Jodrell Bank is one of
the world's largest radio reflectors. When the reflector is pointing in the horizontal direction,
its highest point is 91 metres
above the ground. The instrument is computer-controlled
from the control room in the
main building of the Institute.

64   The pulsar in the Crab
nebula was optically identified.
The same pulse period of 0.033
seconds from this pulsar has
been noted in the radio-frequency, optical and X-ray ranges.
The Crab nebula is composed
of the matter emitted by the
supernova observed in 1054.
The pulsar detected in the Crab
nebula is the neutron star which
survived from the original star
after the supernova outburst.

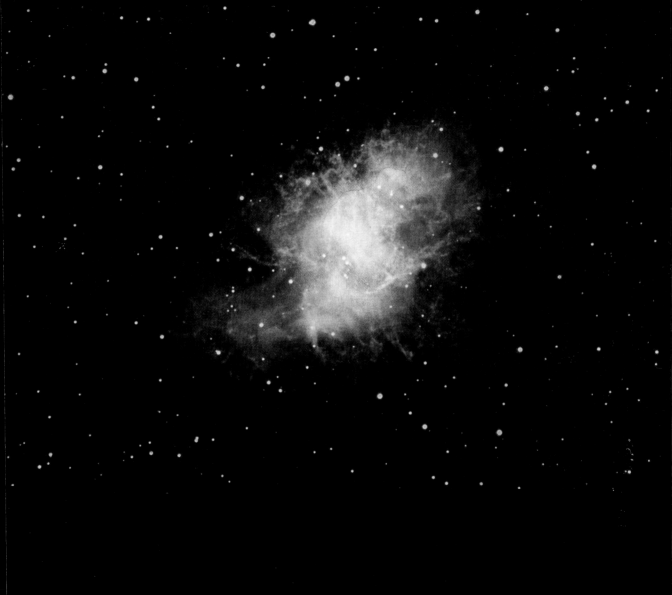

65   The 91-cm telescope of the
G. P. Kuiper Airborne Observatory is installed in a Lockheed
C-141 jet-engined transport aircraft. The square port through
which the telescope has an un-
obstructed view of the sky and
which may be closed off, can be
seen in front of the wings. Since
the instrument is located inside
the fuselage, the direction of
observation is restricted to elevations between 35° and 75° and
in azimuth to 4° on either side
across the direction of flight.

66   Not even the view of this
control console indicates just
how complex is the control and
computer technology of the telescope on board the G. P. Kuiper
Airborne Observatory. The
acquisition and pinpointing of
the objects to be observed and
the necessary directional stabili-
zation of an instrument on board
an aircraft alone necessitate an
exceptional degree of technical
sophistication. The team consists of the three flight crew,
three technicians who work at
the telescope or at the control
console and up to five scientists
who follow the course of the
programme from the scientific
viewpoint and may modify it
if necessary. On the left in the
background, there may be seen
the tailpiece of the telescope to
which detector equipment may
be fitted.

● With a total of 380 staff, the Sternberg State Astronomical Institute is the second largest Soviet research centre in this branch of science. It was founded in 1931 when the Observatory, which had just celebrated its first centenary at the time, was united with the Geodetic Research Institute and the State Astrophysical Institute, both of which had been established by the Soviet Government.

Of the directors of the Observatory, the best-known is F. A. Bredikhin (1831–1904). He was a successful investigator of comets, studying in particular their origin, the development of their outer appearance and their relationship to meteors. Research work on the planetary system also made the name of the director of the Astrophysical Institute, V. G. Fessenkov, well-known in the world of science and elsewhere. He was especially interested in the problem of the formation of the planetary system and also in practical photometric investigations and the development of apparatus.

P. K. Sternberg became the director of the Observatory in 1916 but, immediately after the October Revolution of 1917, was entrusted with important military tasks until his early death in 1920.

For the few remaining members of the staff, the difficult war years were marked by the loss of talented colleagues, evacuation to areas far from the front and a concentration of their work solely on tasks of essential importance at this time, namely the time service and the monitoring of the Sun for the prediction of radio disturbances. Immediately after the war, however, an enthusiastic start was made with the work of reconstruction and since 1956 the Moscow staff of the Sternberg Institute have been housed in the buildings of the Lomonosov University on the Lenin Hills. Since astrophysical observation programmes cannot be carried out in this densely populated area, two outstations are at their disposal.

In the summer of 1957, scientists from the Sternberg Institute organized a mountain expedition in the vicinity of Alma-Ata, the capital of the Kazakh SSR as part of the International Geophysical Year. What was initially of a provisional character only, became a permanent outstation, offering excellent conditions at a height of 3,000 m for the observation of the Sun and faint light phenomena from interplanetary Space.

The second outstation, an observatory in the Crimea, was taken into service in 1958. Since then, its 125-cm reflector has been used for spectroscopic investigations of galaxies, supernovae, gaseous nebulae and variable stars and also for the examination of the planets. Photometric measurements are likewise an additional area of activity.

For the immediate future, a further expansion in research facilities is planned with the establishment of an observatory in Central Asia. Expeditions are already engaged in the collection of data for the assessment of the "astroclimate" in various regions.

The points listed above do not however reveal the entire scope of the scientific work of the Sternberg Institute. This includes radio astronomy, the astrophysics of the galaxies and stars of all types, interstellar matter, the objects of the solar system, the time service and gravimetric work. The Institute has also contributed to modern knowledge of the topography of the Moon. Particularly well known are the *Atlas of the Other Side of the Moon* and the world's first lunar globe, both showing that part of the Earth's satellite which is permanently turned away from us. These two works are based on the photographs transmitted by the Luna 3 automatic station. It is to be noted that the three Chairs of Astronomy at Lomonosov University are held by leading staff of the Sternberg Institute so that the high standards of this great research institute also have a direct influence on the training of young scientists.

● In December 1976, an astronomical news-sheet published the information that the first photographs had been taken with the 3.6-m telescope of the European Southern Observatory (ESO). This was a great leap forward for Western European astronomy, not merely because a new giant telescope had been taken into service but above all since ESO, now equipped as planned, represented a completely new form of research in astronomy. This was the start of a vigorously pursued and clearly defined project, the investigation of the Southern Sky, on the basis of international collaboration with massive technical support and adequate staffing. The ESO is not an observatory in the conventional sense but is rather a modern research organization in the field of astronomy.

The suggestion of building a great observatory in the Southern Hemisphere as a joint undertaking by countries of Western Europe was made by the German astronomer Walter Baade (1893–1960) in 1953. From the early 1930's he had worked with the giant telescopes in California. His proposal was favourably received since it offered a way of drastically reducing the lead of the USA as far as instruments were concerned and of achieving a significant increase in knowledge by the intensive exploration of the Southern Sky which was long overdue. The astronomers concerned had to develop considerable diplomatic activity, since agreement had to be reached on government level. Their efforts were crowned with success in October 1962 when the ESO Convention was signed by the authorized representatives of the countries involved—Belgium, the Federal Republic of Germany, France, the Netherlands and Sweden—in Paris. Denmark joined the Convention in 1967.

The first director-general of ESO from 1962 until his retirement in 1969 was O. Heckmann, who had been in charge of Hamburg Observatory for 20 years prior to this. His scientific reputation and his drive were a major asset to the development of the ESO in the critical early years.

The site of the observatory had to have excellent climatic conditions but, for a wide variety of reasons, also had to be within reach of a fairly large town. The prospecting expeditions for the objective evaluation of the observation conditions initially concentrated on South Africa but the ultimate choice was a location in the central part of Chile at a latitude of about 30° south. The local name for the site is "La Silla"—"the saddle" —describing a slight hollow between the twin peaks of a 2,400-metre mountain. The town of La Serena is relatively near and the capital of Santiago with the ESO administrative centre in Chile is about 600 km away. There is access to the Pan American highway, which is only 40 km away, and to the port of Coquimbo. The latter played an important role as a transshipment point for heavy equipment when ESO was under construction. The Atacama desert close by, the Andes and a cool coastal current provide the atmospheric conditions required. Almost every night of the year can be used for observations and 70 % of the nights are so clear that they are designated by astronomers as "photometric" on account of the clarity and stability of visibility. Air turbulence in the atmosphere is largely absent so that the definition of the star images in the telescope is always very good. To exclude troublesome disturbances in the vicinity for the future as well, an area of 625 sq. km has been purchased from the Government of Chile, a terrain which even includes mountains of over 3,000 m in height.

The European centre of the ESO was initially at Hamburg and it was from there that the organizational work and also the design of the first two telescopes was carried out. In 1970, however, an agreement was made with CERN, the Western European nuclear research centre, which provided for the establishment of a developmental facility for the construction of telescopes and ancillary equipment on the latter's premises in Geneva. The administration of ESO remained at Hamburg for the time being until it ultimately became necessary and possible, with the growth in potential,

11   The diagram depicts the interior of the dome of the 3.6-metre telescope of the European Southern Observatory. The telescope is shown in the coudé arrangement, i. e., the light-path is passed into the fork after being reflected twice by secondary mirrors and then, along the polar axis, arrives at the spectrograph installed on the floor under the dome. However, for more efficient use of the time available, the latter can also be operated in association with an auxiliary telescope with a mirror diameter of 1.5 metres in the neighbouring dome. Other optical components and the prime focus cabin are stored on the left. A high degree of mechanization enables the optical system to be changed with exceptional rapidity so that it is possible, in the course of a single night, to attack observational tasks of quite different types. The slit in the dome is uncovered by moving the slit-cover back over the dome. To avoid wind disturbance from outside, a blind is used to close off the slit up to the observation opening. The adjustment and tracking of the dome and blind when observations are in progress are handled by the control-computer of the telescope. To avoid the layer of turbulence in the vicinity of the ground, the dome building is 25 metres high and the telescope is located on the 5th floor. On the lowest floor, there is, among other equipment, the high-vacuum plant in which an aluminium layer, only a few microns in thickness, is applied to the mirror every two years to ensure a high reflective capability.

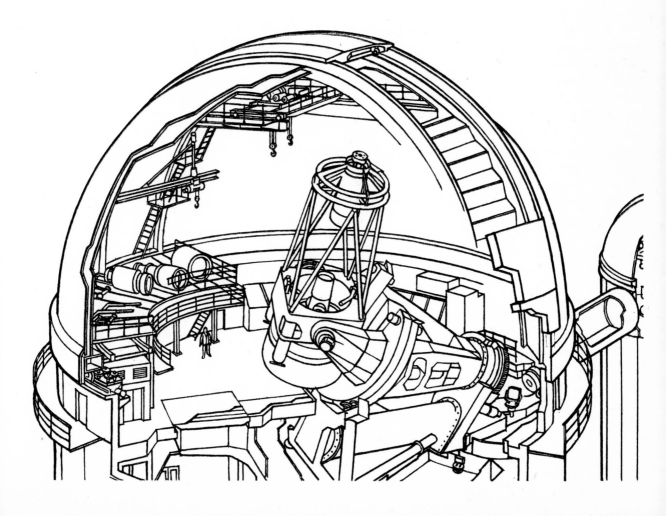

to concentrate all the departments of ESO in Europe at one place. This was how the new European centre came into existence at Garching near Munich in the years after 1977.

The construction of the Observatory on La Silla began in 1965 with the building of an access road. Prefabricated houses for workers and staff employees were then erected, followed by workshops, a simple guest house for all those working at the site for a fairly short period and, of course, the dome buildings. The first telescopes were taken into service at an early stage and, to some extent, even under temporary conditions. This ensured, however, that astronomical results could be published very rapidly, demonstrating the usefulness of the project. At present, ten telescopes are in service on La Silla. The first instrument was a double astrograph from France which can be fitted with an objective prism on one of its two 40-cm telescopes. With the special exposure technique employed, the radial velocities of stars can be deduced from the stellar spectra obtained with the prism. The instrument, which was ESO's largest telescope up to 1976, has a reflector diameter of 1.5 metres and is used exclusively in conjunction with grating spectrographs. The Schmidt telescope with its corrector-plate diameter of 1 metre is of particular importance for the exploration of the Southern Sky. It is being used to obtain pictures for a photographic atlas of the Southern Sky. This is an extension to the southerly zones of a project which was carried out with the "Big Schmidt" of the Hale Observatories in the Northern Sky and which, as the *Palomar Observatory Sky Survey*, has provided a wealth of interesting objects for astronomers. It was inevitable that an urgent wish should be expressed also for a photographic atlas of that part of the sky which could not be covered from California. The work on this programme is being shared by ESO and British astronomers with their Schmidt telescope in Australia. A special photographic laboratory of ESO in Geneva has been given the responsibility for the series production of the atlas in both cases.

The photographic plates from the Schmidt reflector with their wide field of view and great range show a profusion of objects which merit systematic observation. The commissioning of ESO's largest telescope so far, which has a reflector diameter of 3.6 metres, has made this possible. Great versatility in use is guaranteed by the availability of various optical systems, depending on the specific task to be performed. The prime focus is especially suitable for photographic exposures. It is situated in front of the reflector surface and is accessible from the prime focus cabin at the front end of the telescope. It is here that the astronomer on duty sits, in the middle of the tube, as it were, and is moved with the telescope. The spectrographs and photoelectric photometers, however, are arranged in the Cassegrain focus at the back of the pierced principal mirror. Particularly heavy and delicate equipment can be installed underneath the dome area and independently of the moving telescope, the light from celestial objects being brought in through the polar axis with the aid of secondary mirrors. A noteworthy innovation is that the equipment in this "coudé focus", as it is called, can also be used when the 3.6-metre telescope is operating with one of the other systems. The rays from a 1.5-metre telescope located directly next to the great dome are reflected into the system in this case. In this way, test observations or investigations of relatively bright stars can be made without encroaching on the valuable time of the main instrument. At its centre in Western Europe and in Chile, ESO employs about 270 staff from various countries. More than a third of these are technicians, responsible for the operation and further development of the ancillary instruments and equipment. Some 35 scientists are concerned with the development of telescopes, questions of the use of telescopes and especially with matters typical for astronomical research. One third of the total time available on the telescopes is at the disposal of the permanent ESO staff. It is therefore mainly visiting astronomers who come to La Silla, obtain their observation material there with the assistance of the experienced technicians and finally evaluate and publish it at their own institutes. A programme committee decides on the allocation of telescope time on the basis of the scientific importance of the programmes of work submitted. This committee does not have an easy task since the time requested often exceeds that actually available by a factor of three.

● The usual idea of an observatory is a dome structure containing telescopes which are used for the visual, photographic, photoelectric and spectroscopic observation of celestial objects. However, observational data alone cannot immediately supply any new findings.

The data must be interpreted in order to obtain information on the physical structure and chemical composition of stars and stellar systems. Theories on the origin and development of celestial objects are then often derived on the basis of this information. The theoretical considerations, in turn, lead to the systematic observations which are necessary for the confirmation of the theories and result in the acquisition of additional knowledge. There are astronomical institutes which are concerned solely with theory. An example of the many such establishments is the Institute for Astrophysics in the Max Planck Institute for Physics and Astrophysics.

The Max Planck Institute for Physics moved from Göttingen to Munich on 1 September 1958. At the same time, the Institute for Astrophysics was founded as a part of the Max Planck Institute for Physics and Astrophysics. This partial independence for the astrophysics aspect was not a spontaneous act but the result of many years' work on astrophysical problems which had formed an organic part of the activities of the Max Planck Institute for Physics. The Max Planck Institute for Physics had taken the place of the former Kaiser Wilhelm Institute for Physics in 1948. Work had been carried out there on astrophysical problems as early as

the 1930's. It was then that C. F. von Weizsäcker had demonstrated the significance of nuclear physics for the processes concerned in the release of energy in the interior of stars. It is now an accepted astronomical fact that in the interior of stars hydrogen is transmuted to helium at temperatures of more than 10 million K and that a very small fraction of mass is converted into energy when this takes place.

When the scientific programmes of the Institute were re-established after the Second World War, greater priority was given to astrophysical problems than in the past and this was reflected in the establishment of an independent department for astrophysics on 1 July 1947.

In the new department, problems concerned with the origin of the planetary system were examined and the origin and behaviour of cosmic radiation was investigated. Particular importance in this respect was attached to the calculation of the orbits of charged particles in the magnetic field of the Earth. Fundamental work on the behaviour of ionized gases, which often occur in Space, was also carried out during these years. Yet another field of investigation was the propagation of shock waves in greatly rarified gases, as found in stellar atmospheres and in interplanetary and interstellar Space.

In connection with problems of this nature, it was also realized that a constant stream of protons and electrons is emitted from the Sun and that this is responsible, for example, for the development of the tail of a comet. This stream of solar particles is known as the solar wind. In the Astrophysics Department, work was also started on the origin and evolution of stars. The theoretical studies on the development of stars, using the calculation of star models, have enabled very great progress to be made in this area of astrophysics. The starting-point of all calculations was the release of energy process in which helium is produced from hydrogen. With this energy release process, a star with the mass of our Sun can emit energy at a constant level for about ten thousand million years. The greater the mass of a

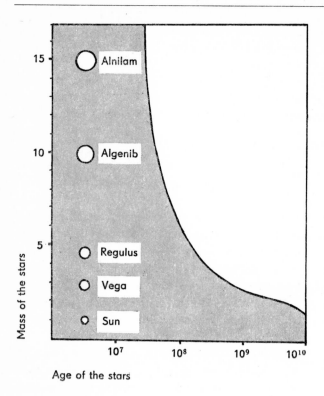

15 - Alnilam

10 - Algenib

5 - Regulus

Vega

Sun

Mass of the stars

10⁷   10⁸   10⁹   10¹⁰

Age of the stars

12 Stars change so slowly that alterations in the course of normal developments can only be observed over a period of time exceeding that of several human generations. Theoretically, it has been deduced that the smaller the mass of a star, the longer its life. The Sun, which is a normal star, will exist for about ten thousand million years. At the moment, it is about five thousand million years old. Stars with a mass ten times that of the Sun, such as Algenib in the constellation Pegasus, will survive for only about 100 million years.

million K and combustion of the helium takes place, carbon then being produced from the helium. The release of energy inside the star leads to the synthesis of heavier and heavier elements. Since these processes cannot be observed, they can only be investigated theoretically. However, the processes in the interior of the star also lead to changes in the outer characteristics, e. g., in brightness, surface temperature, colour and diameter, which can be measured. A comparison between theory and observation is therefore possible.

All the investigations carried out at the Max Planck Institute are of a purely theoretical nature. They are often associated with very detailed calculations and thus call for an advanced level of computer technology. A modern computer installation means the same to a theoretical astrophysicist as a versatile giant telescope to an observing astronomer. This is why, in the Astrophysics Department, a group has been engaged since 1957 with the development and construction of computing systems.

Since the move to Munich, all the work in the now independent Institute for Astrophysics has been carried out in the three departments of Theoretical Astrophysics, Theoretical Plasma Physics and Computer Services.

In the course of time, two further institutes emerged from the Institute for Astrophysics. As early as 1960, experimental work was started on questions of plasma physics and this led to the founding of an Institute for Plasma Physics. The Theoretical Plasma Physics Department of the Institute for Astrophysics was also linked with this new Institute in 1968. Following extensive theoretical work on comets and comet tails at the Max Planck Institute for Physics and Astrophysics, the Institute for Extraterrestrial Physics was established in 1963 and work began on the experimental investigation of man-made comet tails. For this, clouds of gas were released by rockets above the Earth's atmosphere, the further development of which under cosmic influence was observed from the Earth.

star, the more demanding it is with its energy, i. e., the more mass a star has, the quicker its evolution. However, theoretical calculations also indicate what will happen to a star when the whole of the hydrogen in the regions near the centre is converted to helium. The interior of the star contracts and, as a consequence, the temperature at the centre of the star rises to about 100

Since 1 June 1971, the new department for "Relativistic Astrophysics" has been in existence at the Institute for Astrophysics. The staff of this department are concerned with the physics of the neutron stars, black holes, the origin of pulsars, cosmic X-ray radiation and fundamental questions of gravitational theory.

Although slight shifts have taken place on a number of occasions in the research profile of the Institute for Astrophysics, one thing has never changed: the Institute for Astrophysics at the Max Planck Institute for Physics and Astrophysics has remained faithful to the treatment of theoretical problems, i. e., it is the observations of others which are used and interpreted but the theoretical results obtained provide a stimulus for further systematic observations.

In the meantime, more than a hundred staff work at the Institute, including about 45 scientists, and work was started in 1977 on a new building at Munich/Garching for the Institute for Astrophysics. The move to the new building took place in 1980. The Institute for Astrophysics is now in the immediate vicinity of the ESO (European Southern Observatory) building under construction in Garching.

● Since 1970, the two celebrated observatories on Mount Wilson and Palomar Mountain have been known as the Hale Observatories but they have always collaborated closely with each other. They are named in honour of G. E. Hale (1868–1938) who, quite apart from his great scientific achievements, founded the Mount Wilson Observatory and initiated the construction of the 5-metre telescope at Palomar Observatory. The two institutes participate in a joint programme of astronomical research under a single director but continue to have independent administrations. Mount Wilson Observatory is sponsored by the Carnegie Institution of Washington while the observatory on Palomar Mountain is responsible to the California Institute of Technology. Other research centres operated by the Hale Observatories are Big Bear Solar Observatory and Las Campanas Observatory in the Chilean Sierra del Condor.

With the observatory at Las Campanas in the Southern Hemisphere, a project was completed in 1976 which had been envisaged by G. E. Hale at the time that he founded Mount Wilson Observatory but which he had been prevented from implementing by financial difficulties. The principal instrument on the 2,300-metre peak in the Sierra del Condor is a 2.5-metre telescope. It is named after Irénée du Pont, an industrialist who took an interest in science, and it was his daughter who generously endowed the trust which enabled the telescope to be constructed.

Solar research was an important element in the motivation which led to the founding of Mount Wilson Observatory and this branch of astronomy has been a significant element in the research profile of the Observatory for its entire history. The appropriate observational facilities were expanded in 1969 by the commissioning of a battery of telescopes on a common mounting for studying the fine structure of the solar atmosphere. To ensure the best atmospheric observational conditions, the Observatory was built as a tower right in the middle of the Big Bear Lake. The large area of water surround-

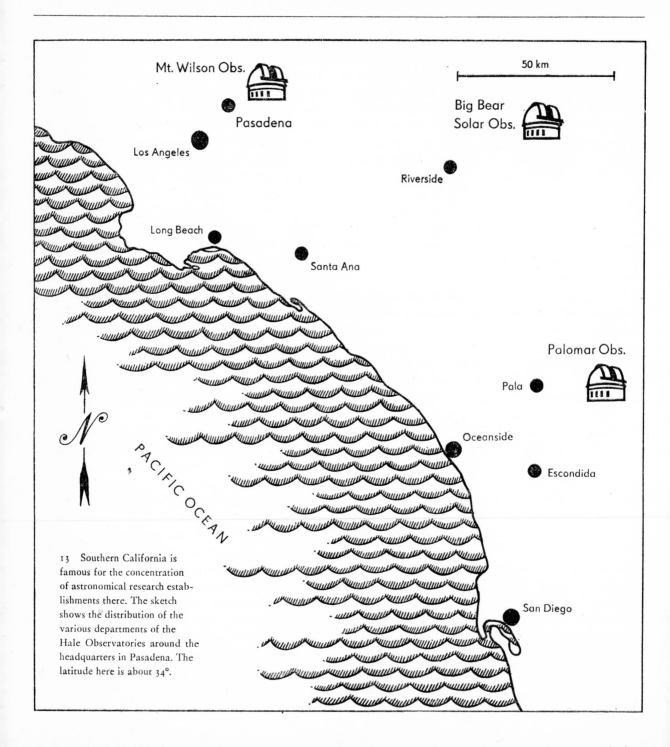

Mt. Wilson Obs.

Pasadena

Los Angeles

50 km

Big Bear
Solar Obs.

Riverside

Long Beach

Santa Ana

PACIFIC OCEAN

N

Palomar Obs.

Pala

Oceanside

Escondida

San Diego

13   Southern California is
famous for the concentration
of astronomical research establishments there. The sketch
shows the distribution of the
various departments of the
Hale Observatories around the
headquarters in Pasadena. The
latitude here is about 34°.

ing the tower avoids the troublesome air turbulence which is found at many solar observatories at about mid-day as a consequence of the build-up of heat in the ground.

The headquarters of the Hale Observatories are located at Pasadena. Apart from the telescopes, there is everything there which is necessary for the scientific work, including various laboratories and workshops for the further technical development of the research instruments. The optical workshop of the Hale Observatories, where large telescope reflectors of high quality —including the 5-metre reflector—have been ground, is especially famous.

It may be stated without exaggeration that the Hale Observatories have left their mark on modern astronomy in a manner matched by no other institution. The annual reports list papers on the monitoring and investigation of the Sun and of its magnetic field, they mention the successful search for small planets and new satellites of the major planets and there are reports about spectroscopic and photometric projects carried out with modern detectors and referring to almost all types of stars, to star clusters and stellar systems, to interstellar matter, and gas nebulae. Questions concerning the physical and chemical state, the spatial structure and the development of individual objects and Space as a whole are of central importance. About 150 scientific papers by staff and visiting astronomers at the Institute are published every year. Every night, even if it is free of cloud for only part of the time, is used to acquire new observational material. This means that about 2,500 hours are spent in nocturnal observations per year. In addition, the 5-metre Hale Telescope is also used in twilight conditions to carry out measurement tasks which do not call for absolute darkness. These involve certain investigations of the bright planets and measurements in the infrared spectral area and on wavelengths of about 1 mm, an interesting area which represents the transition from remote infrared to the radio waves.

Of the outstanding personalities in the history of the Observatories and in addition to G. E. Hale, mention may be made by way of example of W. S. Adams with his contributions to stellar spectroscopy, A. A. Michelson who measured star diameters with an interferometer arrangement on the 2.5-metre telescope and, above all, E. Hubble (1889–1953) and W. Baade (1893–1960) who made full use of the power of the telescope to penetrate far into the depths of Space. A brief outline of the historical development of the two observatories is given below.

● The Carnegie Institution of Washington was founded in 1902 and endowed with ten million dollars by A. Carnegie. Its aim was defined as the generous promotion of research and of the application of scientific findings for the benefit of mankind. After difficult negotiations with the committees entrusted with the use of the financial resources and in the face of opposition from conservative astronomers, G. E. Hale obtained an unheard-of sum of money for the erection of a new observatory on Mount Wilson in Southern California. This fund was sufficient to implement the major part of his ambitious plan—a solar observatory with a 1.5-metre reflector for astrophysical research. In view of this success, the abandonment of his project for an observatory in the Southern Hemisphere was not quite such a blow. G. E. Hale also played a leading part in the founding of the Yerkes Observatory in the East of the United States and, as its director, contributed to its development. He had always strived to find a climatically suitable place for solar observations. Furthermore, he advocated the construction of a giant reflecting telescope since it was his view—which has been fully confirmed in the meantime—that the future of observational astronomy at that time was associated with this type of telescope. For G. E. Hale, the significance of the expansion of observational possibilities with a giant telescope resulted from the long since necessary inclusion of astrophysical thinking in astronomy. He justified the unusual combination of a solar observatory with an instrument designed for stellar astronomy by the sentence: "The story of the origin of the sun and its development is illustrated in stars of many types which are no less important to a thorough understanding of its physical constitution than is a direct investigation of solar phenomena."

Earlier expeditions had already identified Mount Wilson (1,742 metres) as a particularly suitable site for astronomical observations. However, it was only under G. E. Hale that a solar telescope with a horizontal optical path was erected in 1904 as the first instrument of the new Mount Wilson Observatory. The establishment of the mountain was associated with considerable difficulties since the only access was along a fairly unsafe rocky path suitable only for mules. The next instrument to be installed on Mount Wilson was a tower telescope with a focal length of 20 metres, which was followed in 1910 by a 45-metre instrument. The concept of the tower telescope as such was evolved by G. E. Hale.

Work began on the great reflector in 1904, the year when Mount Wilson Observatory was founded. The basic element of this was a glass block of somewhat more than 1.5 metres in diameter which G. E. Hale had received from his father. This glass blank was ground in an optical workshop at the headquarters in Pasadena and the telescope was taken into service in 1908. Powerful though the instrument was, the question of the structure of our more distant cosmic environment, the regime of the faint stars, called for the use of even larger telescope apertures. Once again, it was a private donation which enabled progress to be achieved and the 2.5-metre Hooker Telescope, the "100-inch", was commissioned in 1917 after delays caused by the war. For 30 years, it remained the largest and by far the most powerful telescope in the world. It marked the final swing in favour of astrophysics and especially the study of stellar systems at Mount Wilson. Today, the Observatory is greatly inconvenienced by the brightness of the night sky resulting from the almost total urbanization of the coastal strip around Los Angeles. This has meant that observations have had to be restricted to projects which are less dependent on background brightness.

● The scientific problems raised by the work carried out with the two great telescopes on Mount Wilson very soon convinced G. E. Hale and his colleagues that there was a need for an even bigger research instrument. With his usual energy, G. E. Hale threw himself into the work of the technical preparation involved and the problem of raising funds. By 1928, he had obtained a promise of 6 million dollars with which the famous 5-metre telescope was subsequently built—apart from an additional $ 500,000 which had to be raised after the war to cover increased costs.

Palomar Mountain, 1,706 metres high, had already been considered as a possible site for what subsequently became Mount Wilson Observatory but the idea had been dropped at the time on account of its inaccessible location. It was chosen as the site for the new giant telescope since it was screened from the conurbation of Los Angeles not only by distance but also by the ranges of hills between and even today the night sky is still relatively dark.

On account of its low thermal expansion, Pyrex glass was selected as the material for the great glass block of the 5-metre reflector. To save mass, it was decided to take a new technical approach by which the mirror was to consist of a relatively thin disc, reinforced by ribs at the back. The mirror surface at the front is only 10 cm thick while the total thickness at the ribs is 60 cm. An additional mechanical support of the mirror in its mounting then supplies the stability required. The casting of the disc only succeeded at the second attempt.

In December 1934, 20 tons of molten Pyrex was poured into the special mould after which it was kept for two months at the melt-temperature and then allowed to cool in the course of a further ten months. After an adventurous journey right across the continent on a special transporter, the blank arrived at Pasadena in 1936. The process of grinding it then began and continued, with an interruption of four years during the Second World War, for eleven years. Five tons of glass were removed from the huge disc by a meticulous and protracted process of grinding and polishing. The first "light" struck the reflector in December 1947 when it was installed in the mounting which had been built and tested in the meantime. Corrections to the optical surface were then carried out and regular observations finally commenced two years after the first tests of the optical system had been made. G. E. Hale did not live to see the crowning of his life's work but it is in his honour that this powerful reflecting telescope bears his name.

The Hale Telescope makes an awe-inspiring impression not only on astronomers but also on the almost 100,000 visitors who come every year to the glass-enclosed viewing-gallery. With a diameter of 6.5 metres and a length of 18 metres, the main tube is massive in appearance. The size of the instrument is evident from the fact that the mirror has a hole of one metre in diameter at its centre for its Cassegrain focus. This is large enough for the entire tube of a 90-cm telescope to pass through which is itself big enough to be included in the medium-size class of instruments! The total moving weight is 530 tons but oil-pressure bearings and perfect balance mean that only low-power motors are required for the pointing and tracking movements. Various optical systems can be used in the prime focus, to which there is access from a cabin carried on the front end of the tube, in the Cassegrain focus and in the coudé focus. The 1,000-ton dome, which is turned automatically with the telescope, is likewise of an impressive size. Its highest point is 30 metres above the floor of the dome, thus being equivalent to the height of a twelve-storey building.

In addition to the Hale Telescope, there is another instrument on Palomar Mountain which has achieved fame: the "Big Schmidt"—a Schmidt camera with a 1.2-metre unobscured aperture and a 1.8-metre mirror diameter. Whereas the optical system of the 5-metre telescope is primarily designed for the examination of individual objects, the "Big Schmidt" with a field of view equivalent to the apparent area of 150 full moons enables "panorama" photographs to be taken for the study of cosmic objects in abundance. To make such material accessible to a large number of astronomers, the first comprehensive observational programme was the project for a photographic atlas of the sky and, after seven years of work, this was completed in 1959. This *Palomar Observatory Sky Survey*, as it is known, consists of 935 pairs of black-and-white negative glass plates which show the entire sky north of the declination – 30° in the light of two wavelength ranges, red and blue. Sets of paper copies are sold to all observatories and they represent an enormous fund of knowledge, much used throughout the world, for all possible investigations concerning the classification, distribution, position determination and even photometry of the various kinds of cosmic bodies.

● Pulkovo Observatory, near Leningrad, is one of the best-known and oldest observatories in the world and, in earlier times, it was often called the "astronomical capital of the world". In its development to date, the Institute has passed through several decisive phases. On the instructions of Czar Peter I, the "Art Cabinet" building was erected on the northern bank of the Neva between 1718 and 1734. It contained an "art and curiosity cabinet" and, even at that stage, an observatory. The Academy of Sciences assumed responsibility for the Observatory founded in 1718 and for all the following observatories in the Leningrad area. In accordance with the custom of the time, the Observatory was accommodated in a tower in the central part of the building. Like the entire building, the Observatory was laid out as a museum in the main. The greatest Russian scholar of this phase of the Observatory was M. V. Lomonosov who concluded from observations of Venus as it passed across the Sun that the former must have an atmosphere.

Like many monarchs of the time, Nicholas I wanted a fine observatory in his realm also and he was not satisfied with the one founded in 1718. Consequently, the foundation-stone for the present-day Pulkovo Main Astronomical Observatory was laid in 1835 in the neighbourhood of Leningrad. The terrain is on a hill 15 km to the south of the city in the immediate vicinity of the Czar's summer residence. The land was a gift from the Czar and the Observatory is often called the "Nikolai Central Observatory" after him.

The first fifty years of the second phase of development are intimately associated with the name of F. G. W. Struve. Six important astronomers came from the Struve family and their influence is still felt. F. G. W. Struve was the director of Pulkovo Observatory from 1839 to 1862. He also gave the Institute the scientific programme which inspired research work at Pulkovo long after he had departed. Scientific research was concentrated on positional astronomy and was closely linked with the geodetic and geographical work carried out in Russia. Even today, the headquarters of all geodetic surveys of the USSR is in the main building at Pulkovo. F. G. W. Struve was the founder of an entire scientific school of astronomy. He and his staff succeeded in combining the "art of observation" with the "science of observation". His observations and research work in the field of the double stars are regarded, even today, as some of the most important publications in this area of astronomy. On the basis of improved observations, he also re-determined the aberration constant. At the same time as F. W. Bessel and T. Henderson, he succeeded in measuring the stellar parallax of the star Vega in the constellation Lyra.

His son, O. Struve, followed him as director of the Observatory from 1862 to 1889. Sons of O. Struve were directors of the observatories at Kharkov and Berlin-Babelsberg. Another member of the Struve family was director of the American radio observatory at Green Bank after the Second World War.

When the Observatory was enlarged in 1839, the principal instrument at Pulkovo was a refractor with an aperture of 38 cm. The lens for this telescope was ground by the celebrated optician J. von Fraunhofer. During the time of O. Struve, the Observatory acquired a 76-cm refractor with a focal length of 14 metres. Optical characteristics of this order made it one of the biggest refractors of the world. Observations with this instrument commenced in 1885. A double astrograph with a photographic lens of 33 cm and a visual lens of 25 cm was erected in 1893. The "Bredikhin" astrograph, as it is known, was presented to Pulkovo Observatory by the famous astronomer F. A. Bredikhin at the turn of the century. Systematic observations of latitude began in 1904 with a 13.5-cm zenith telescope. These observations were only interrupted by the Second World War and have been continued since 1948. In 1915, a 1-metre reflector and an 82-cm refractor were ordered for the Pulkovo Main Observatory. The reflecting telescope was subsequently erected in 1925 at the outstation of Pulkovo Observatory in the Crimea. This instrument was destroyed during the Second World War. The refractor which had been ordered was never installed although, in renewed negotiations with the manufacturers, it had even been anticipated that a lens of 104 cm diameter should be ground instead of one of 82 cm.

The importance of outstations in climatically favourable regions was appreciated at an early stage at Pulkovo and observations in the Crimea and the development of Simeiz Observatory began as early as 1908.

The first astronomical investigations at Pulkovo Observatory were concerned with the Sun. From observations during the eclipse of the Sun, O. Struve showed that the prominences and the solar corona are not optical effects, as assumed hitherto, but parts of the Sun itself. This was followed by brightness measurements of stars and colours were determined for all the stars in the *Bonner Durchmusterung* in 91 Kapteyn areas.

The beginning of spectroscopic work and radial velocity determination in Russia is closely associated with the name of A. A. Belopolsky, an astronomer of Pulkovo. Despite the growing importance of astrophysical problems, the astrometric questions which had dominated the work of the Observatory since the time of F. G. W. Struve were still pursued in a systematic manner. In 1932, for instance, a catalogue was published with the exact positions and proper motions of 20,000 stars.

The Second World War interrupted the work of Pulkovo Observatory and led to its complete destruction. It was in the front line during the siege of Leningrad. The precious lenses and many instruments were moved to Leningrad and thus the 76-cm lens of the refractor was saved although the actual instrument was destroyed. As early as 11 March 1945, the Government of the USSR took the decision to rebuild the Central Astronomical Observatory. The main building has since been re-erected in the historical style according to the old plans.

It was not long before Pulkovo Observatory was once more one of the leading observatories of the USSR. Today, work is being carried out on the following scientific problems there:

- fundamental astrometry,
- astronomical constants and the motion of the Earth's poles,
- time service,
- problems of stellar physics,
- solar physics,
- radio astronomy,
- construction of astronomical instruments.

In particular, the Instrument Construction Department at Pulkovo is of great importance for all astronomical activities in the USSR. It was at Pulkovo that the blueprints were prepared for the 6-metre telescope at Zelenchukskaya, the world's largest.

Haute-Provence Observatory
Saint-Michel, France

Lick Observatory, University of California
San Jose, California, USA

● Haute-Provence Observatory is a national observational institution in the truest sense of the word. The decision to create such an institute in France was taken in 1936, at a time when I. Joliot-Curie was the First Deputy Secretary of State for Scientific Research. Since French astronomers had long wished to have a well-equipped observational station, five years had already been spent in searching for a suitable site. The choice fell on an area about 100 km to the north of Marseilles, on a plateau of 650 metres above sea level at the very foot of the Alps. It was calculated that observations could be carried out on 250 nights per year and on 100 of these for the full night.

The rapidity with which work was initially commenced on the Observatory was interrupted by the Second World War. Nevertheless, a 1.2-metre telescope was taken into service in 1942. The Institute is now one of the world's most comprehensively equipped observing stations and there are 13 domes on the terrain of the Observatory which is 1 sq. km in area.

The major instruments are:

1.93-metre telescope with coudé and Cassegrain systems and prime focus, in service since 1958,
1.52-metre telescope with a coudé system, in service since 1969,
1.2-metre telescope, in service since 1942,
plus 1-metre, 80-cm, 60-cm and 40-cm telescopes and a 60/90-cm Schmidt camera.

The telescopes of Haute-Provence are equipped with superb spectrographs, electronic cameras, infrared detectors and photometers. Nearly all of these important ancillary instruments were developed and constructed in the Institute's own workshops.

Special astronomical research programmes, which are typical of many other observatories, cannot be quoted for Haute-Provence Observatory since it has been organized as a national observational institute where visiting astronomers, often from other countries, work for fairly short periods. The observational tasks are determined by the problems of astronomers at the other French observatories and foreign institutes. The work profile of Haute-Provence Observatory also determines the structure of the permanent staff, which is likewise very different from that of traditional observatories. About 90 % of the staff are engineers and technicians. Because of the comprehensive experience in the development of instruments and equipment available at Haute-Provence Observatory, it was only natural that a major contribution to the construction of the 3.6-metre telescope sponsored jointly by France, Canada and Hawaii should have been made there.

● Lick Observatory is one of the oldest, permanently occupied mountain observatories. Observations commenced in 1888 on Mount Hamilton, not far from the town of San Jose. The Observatory bears the name of the eccentric millionaire J. Lick who donated three million dollars in 1874 to a Board of Trustees. His provisions with respect to the Observatory were: "To expend the sum of seven hundred thousand dollars . . . for the purpose of purchasing the land, and constructing and putting up on such land . . . a powerful telescope, superior to and more powerful than any telescope yet made . . . and also a suitable Observatory connected therewith". J. Lick further provided that the observatory should be assigned to the University of California as its Astronomical Department.

J. Lick took a personal interest in the selection of the site for the observatory and initially investigated Lake Tahoe and Mount St. Helena as possible locations. Mount Hamilton, 1,283 metres in height, was the final choice. The survey on this mountain was carried out by T. E. Fraser, a close associate of J. Lick. In 1875, J. Lick accepted Fraser's recommendation on the condition that a road be built from Santa Clara to the summit of Mount Hamilton. This road to the site of the observatory was completed in the autumn of 1876.

J. Lick died at the age of 80 years on 1 October 1876. In life he never ascended Mount Hamilton but it was well within sight of his estate near Santa Clara. However, his last resting place is in the base of the 90-cm telescope, the world's largest telescope at the time it was built. A simple bronze plaque with the inscription "Here lies the body of James Lick" still draws attention to this unique tomb. However, it was twelve years after his death that his body was placed there—where it still remains—since work on the 90-cm refractor was completed only in 1888, the Institute being officially handed over to the University of California on 1 July of that year. The headquarters of Lick Observatory are now on the University of California campus in Santa Cruz in a new building dating from 1966 which also contains study facilities for astronomers, a very large library for astronomy and mathematics, laboratory facilities for measuring instruments, optical and mechanical workshops, a photographic laboratory and a computing department. The main building on Mount Hamilton has two domes. One of them accommodates the 90-cm refractor of 1888, the other a 30-cm refractor. The mounting for this telescope was supplied by the well-known firm of Warner and Swasey. The dome, 90 tons in weight, was manufactured for the refractor by an ironworks in San Francisco. The 90-cm refractor is still used on every clear night for observations, especially of the Moon, the planets and double stars. Lick astronomers have discovered more than 5,000 pairs of double stars with this telescope. The radial velocities of about 2,000 stars have been calculated from spectrograms obtained with the aid of the 90-cm refractor, enabling the motion of the Sun to be determined in relation to neighbouring stellar systems.

When it became possible, about 1880, to make accurate concave mirrors, a 90-cm mirror was made by A. A. Common who sold it to E. Crossley of Halifax, England. Since the astronomical observing conditions there were not good, the latter presented the telescope to Lick Observatory in 1895. J. E. Keeler used it for a

brilliant series of photographs of galaxies. Following the use of the Crossley Reflector on Mount Hamilton for this work, it was said that "the reflecting telescope was born again". Up to this time, refractors had dominated observational astronomy although large reflectors had been constructed by W. Herschel. The results obtained with the Crossley Telescope provided a major stimulus for the construction of larger reflecting telescopes and it was now that this type of instrument began its triumphal progress.

The 3-metre instrument of Lick Observatory with which the first observations were made in 1959 is one of the giant reflecting telescopes which are especially characteristic of the second half of the 20th century. The total cost of this telescope was about four times greater than James Lick's original gift for the founding of the Observatory on Mount Hamilton. The Pyrex glass blank for the 3-metre reflector was cast in 1933 by the Corning Glass Company in connection with the making of the 5-metre reflector. The 3-metre blank was intended as a test flat for figuring the 5-metre mirror but was never used for this purpose. In 1951, the blank and grinding equipment were brought to Mount Hamilton where eight years of careful work were necessary to grind the paraboloid shape required for the telescope. The greatest depth at the centre of the mirror is about 3.5 cm and a total of 325 kg of glass had to be ground away. The weight of the finished mirror is 4.5 tons which is relatively low for a mirror of this size. Like the larger one, the 3-metre mirror is a thin structure which owes its stiffness and thus its dimensional stability to a system of ribs supporting it.

When the instrument was commissioned in 1959, the prime focus and the coudé focus were available for astronomical work. The prime focus is used for direct photography and is equipped with a photographic spectrograph and a photoelectric photometer. A high-resolution spectrograph is permanently available in the coudé focus. In 1969, the Cassegrain focus was also taken into service, enabling work to be carried out with dispersions between those of the spectrographs at the prime and coudé foci. Data acquisition in the Cassegrain focus is fully automated. All information acquired with the instruments in the Cassegrain focus is sent to a data room where it is immediately processed by computers. There is also a remote-control facility for the telescope from the data room. For the first time in astronomical history, astronomers could see the spectrum of a star, for example, while it is being recorded. One of the most spectacular series of observations with the 3-metre telescope began after the Apollo 11 mission. For the first time, a laser beam was aimed at the Moon, was reflected back to Earth by a laser reflector set up by the Apollo 11 astronauts and was picked up there by the telescope. The distance between the Earth and the Moon could then be calculated with a high degree of accuracy from the time taken by the laser pulse.

To make full use of the high-resolution coudé spectrograph on all clear nights, a 60-cm telescope was set up in the direct vicinity of the 3-metre telescope on the south side. This telescope is used with the coudé spectrograph when the 3-metre reflector is being utilized via the prime or Cassegrain foci.

Since its early days, Lick Observatory has done pioneer work in many fields of astronomy. This includes the double-star observations of E. E. Barnard, W. Hussey and R. G. Aitken. The fifth moon of Jupiter, the first after Galileo's discoveries, was observed by E. E. Barnard in 1892 and, in all, four of Jupiter's moons were discovered by astronomers working at Lick Observatory. At the end of the 19th century, W. W. Campbell began his historic spectroscopic investigations and radial velocity determinations on the basis of which he and J. H. Moore elaborated the very precise *Lick System of Radial Velocities*. R. J. Trumpler gave the first convincing proof of the widespread absorption and reddening of starlight by interstellar particles. A. E. Whitford subsequently determined the spectral characteristics of interstellar extinction and established the *Whitford Interstellar Extinction Curve*. The Galaxy Counts

of C. D. Shane and C. A. Wirtanen date from the 1950's and 1960's. Today, at Lick Observatory, a staff of one hundred scientific and technical workers are studying problems of the galaxies and quasars and the structure, composition and evolution of stars and star clusters. Since the brightness of the lights of San Jose is proving increasingly detrimental to night-time observations at Mount Hamilton Observatory, there are plans for an outstation to be established at a more favourable site in the future, possibly on Junipero Sierra Peak in California, and to be equipped with an instrument of the 2-metre to 3-metre class.

67   The Sternberg State
Astronomical Institute is on
the campus of the Lomonosov
University at Moscow on the
Lenin Hills. The instruments in
the scattered domes are largely
used for the training of students,
almost all the scientific observations being made at the outstation.

Page 126:
68   The principal instrument
of the Crimean Station of the
Sternberg State Astronomical
Institute is a 1.25-metre telescope which can be converted
to various optical systems with
the aid of auxiliary mirrors. The
picture shows the telescope in
its Cassegrain version with the
measuring instruments—in this
case a spectrograph—arranged
behind the reflector.

69   The illustration shows a
zenith telescope of the APM-2
type which is used at the Sternberg State Astronomical Institute for monitoring the variations of the altitude of the pole.
It can only be employed for
the observation of stars close to
the zenith but then gives such
accurate data for the geographical latitude of the point of
observation at that moment that
it is possible to follow the slow
wandering of the Earth's rotational pole which takes place
within a circle of only some
20 metres in diameter.

70   The 3.6-metre telescope of the European Southern Observatory was constructed under the technical direction of the telescope-development group based at the CERN Western European Nuclear Research Centre in collaboration with a whole series of manufacturers from the most diverse branches. The contract for the general construction and the final erection in Chile was given to a French heavy engineering company. The blank for the main mirror was made in the USA from 13 pieces of quartz. Quartz as a mirror material has now given way to new materials but 20 years ago it was still the material with the lowest thermal expansion so that the exact shape of the mirror and thus the quality of the optical image was retained even with falling temperatures in the course of the night. The actual "barrel" is arranged as a skeleton tube with the cage-shaped cabin of the Cassegrain focus located directly behind the mirror mounting. The secondary mirrors for the Cassegrain and coudé arrangements are at the front end of the instrument where the cabin for observers working at the prime

focus may also be located, as shown in the photograph. The tube of the telescope can be moved "in elevation" around the declination axis which is pointing here in the direction of the slit in the dome. For the setting of the other stellar coordinate, the right ascension, or in order to follow the daily rotation of the sky, the telescope is turned around the polar axis which points upwards to the southern celestial pole. The upper bearing of this axis is nine metres in diameter, is shaped like a horseshoe and is carried on a cushion of oil under pressure. At the left edge of the picture, it is just possible to see the windows of the control room where the staff work who operate the instrument.

71   The western spurs of the Andes in Central Chile offer excellent conditions for astronomical work and this is why several astronomical research centres have been established there in the last 15 years. This aerial view shows the European Southern Observatory on La Silla at 2,400 metres above sea level. The landscape of brown, weathered rock with scarcely any vegetation is an indication of the aridity of the area which is an advantage for astronomical observations. The residential and service buildings necessary for the operation of an observatory in a remote area, the study facilities and the dome buildings are situated at some distance from each other. A dominating feature is the dome of the 3.6-metre telescope

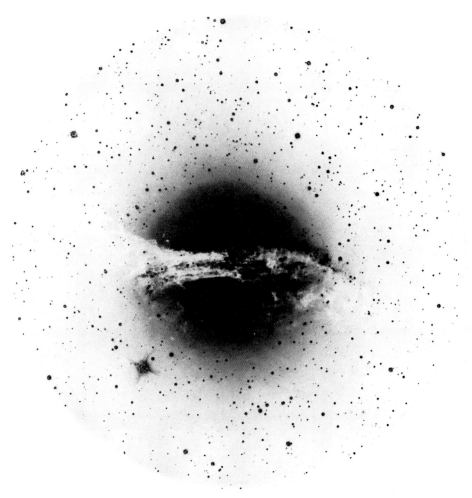

with its diameter of 30 metres. The tower-like dome building of the 1.5-metre ancillary telescope can be seen directly next to it.

72   Centaurus A is 12.5 million light-years away from us and is known as one of the most powerful radio sources in the sky. With the great 3.6-metre telescope of the European Southern Observatory, the characteristics of this remarkable extragalactic stellar system in the visible spectral region can be investigated in a particularly effective manner. In front of this system lit by the light of unresolved stars, there is a belt of light-absorbing matter in the form of dust. The individual stars scattered over the entire photographic plate are foreground objects which are part of our closer environment. In this photograph taken at the prime focus, the diameter of the field is about 1°, which is equivalent to two full-moon diameters. The cross-shaped diffraction figures occurring at the brightest stars are image disturbances at the cross-member supporting the prime focus cabin.

73   This veteran among the instruments of the Hale Observatories has been in the service of astronomical research for 70 years: the 1.5-metre telescope on Mount Wilson.

74   Parts of the 5-metre Hale Telescope on Palomar Mountain can be seen through the open slit of the 40-metre dome.

Pages 132/133:
75   The observer at the reflector of the 5-metre Hale Telescope on Palomar Mountain can scarcely be seen against the mighty dimensions of the instrument. This impressive view is underlined by the massive mounting of the telescope which must guarantee stability and precision of movement.

76   The first test observations with the 5-metre Hale Telescope

on Palomar Mountain indicated that fine corrections to the surface of the mirror would have to be made. These were carried out in 1948 and 1949 with a small polishing-machine in the dome itself to avoid returning the mirror to the workshop in Pasadena. Since the reflecting aluminium coating had not then been evaporated on the mirror surface, the honeycomb structure of the mighty glass block can be recognized on the inside.

77   On the front end of the 5-metre Hale Telescope, there is a cylindrical room which is occupied by the observer when working at the prime focus. He moves with the telescope at all times. Due to the perspective, the mirror appears rather small in size but in actual fact the observer cabin obscures only about 10 % of the area of the mirror.

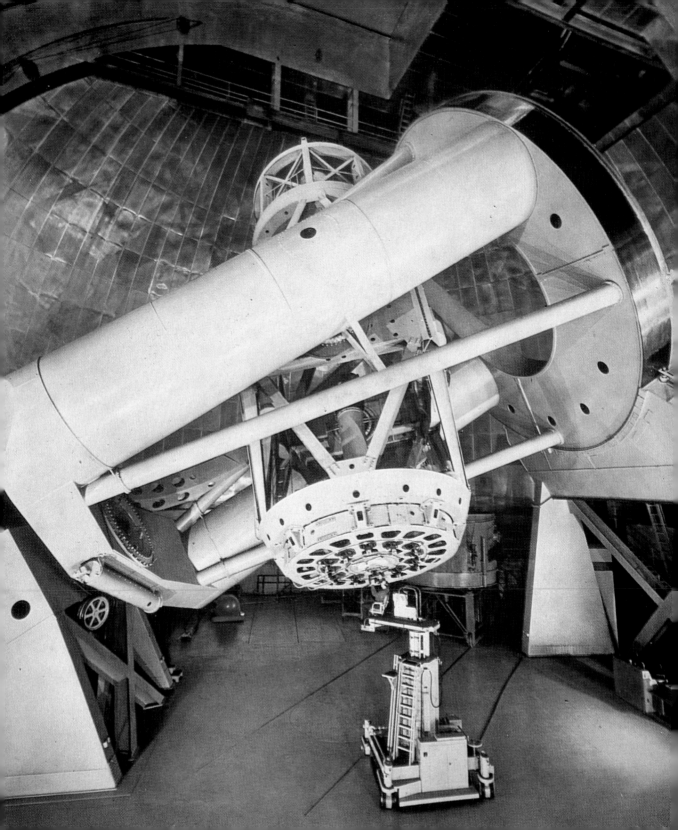

78   The photograph shows the main building of Pulkovo Observatory with its three domes. Other domes can also be seen on the terrain of the Observatory, these dating from the reconstruction after the Second World War.

79   F. G. W. Struve was born at Altona in 1793 and began his studies at Dorpat University in 1808, obtaining his doctorate there in 1813. In the same year, he was given the position of observer at Dorpat Observatory and thus became responsible for this establishment. Already at this stage, he concentrated his attention on the observation of double stars. His first great double-star catalogue, containing the data of 3,112 double stars, appeared in 1827.

Mensurae micrometricae stellarum duplicium etc., listing the measurements of 2,709 double stars and constituting the basic publication in doublestar astronomy, was printed in 1837. The third part of this great project was published in 1852. The catalogue contained information on the right ascension and declination of 2,874 stars, most of which were double stars. By this time, F. G. W. Struve had already become director of Pulkovo Observatory. In addition to his extensive purely astronomical work, he also completed the great Russian grid survey, covering 25 degrees of latitude. Following a serious illness, he entrusted the management of Pulkovo Observatory to his son O. Struve in 1862. F. G. W. Struve died in 1864.

80 The terrain of Haute-Provence Observatory is 1 sq. km
in area. The administration and
laboratories are housed in the
right-hand building in the centre
of the picture while on the left
there are the mechanical workshops. In the largest visible
dome, there is the 1.93-metre
telescope while the oldest instrument, the 1.2-metre reflector,
is accommodated in the dome
on the right of the photograph.
Domes of other instruments can
be seen in the background.

Page 136:
81 The 1.93-metre telescope
is housed in a 20-metre dome.
The instrument has an English
mounting and the total moving
weight is 70 tons. The mirror of
the telescope is relatively thin
and thus weighs only 1.2 tons.

82 The 1-metre Marly Telescope of Haute-Provence Observatory.

83 The main building of Lick Observatory can be identified just below the centre of this aerial photograph. In the large dome on the right of the building, there is the 90-cm refractor and in the left dome the 30-cm refractor. The dome of the Crossley Reflector can be seen on the lower right of the picture. The 3-metre reflector is housed in the great dome in the centre of the photograph.

Pages 138/139:
84 On this picture of Lick astronomers of 1922, J. H. Moore, F. J. Neubauer, W. J. Luyten, R. H. Tucker, W. W. Campbell, R. H. Baker, W. H. Wright, R. G. Aitken and J. A. Pearce (left to right) are standing below the 90-cm refractor. Mrs. A. G. Marshall (secretary), Miss M. Howe and Miss M. Powell are seated in the front row.

85 The photograph was taken shortly after the first aluminium coating had been applied to the 3-metre mirror. The flawless coating of aluminium on the glass is 0.0001 mm thick. When carefully protected from dust and atmospheric pollution, the coating lasts for five to ten years.

86 The 3-metre telescope has an open, skeleton tube. The

"tube" in its fork mounting weighs 45 tons. So that the total moving weight of the telescope, amounting to 150 tons, can be moved easily, rapidly and, above all, with precision, it "floats" on a cushion of oil under pressure.

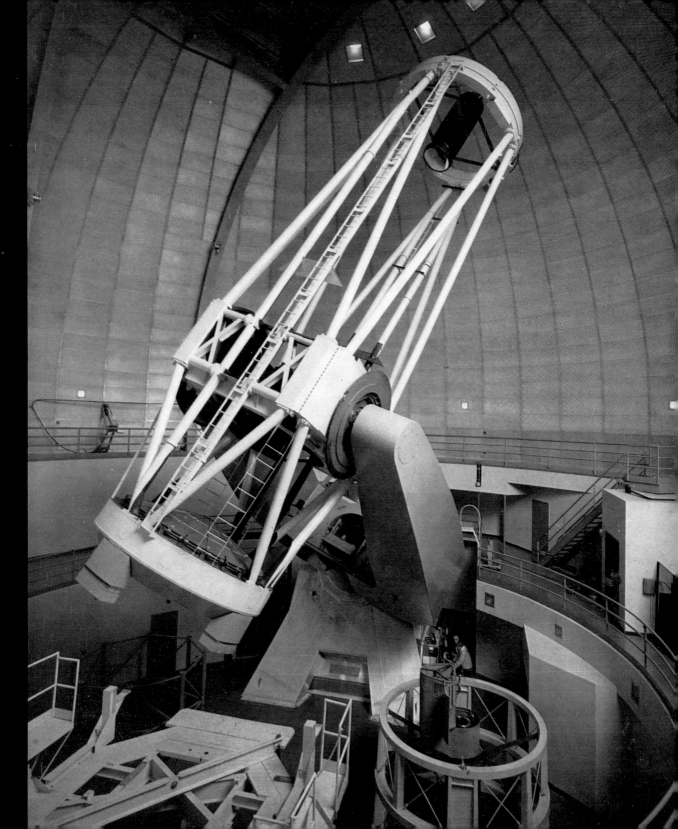

87   Even after more than ten
years, the aluminium-sheathed
dome of the 2-metre telescope
is still a brilliant sight. This is
clear evidence of the purity
of the air in the region of Shemakha Astrophysical Observatory.

● Shemakha Astrophysical Observatory is an institute of the Academy of Sciences of the Azerbaidzhan SSR and its history began in March 1953 when an expedition was organized by the Institute for Physics and Mathematics of the Academy of Sciences. The expedition was headed by G. F. Sultanov, now a member of the Academy and director of the Observatory, and his task was to find a site for an astronomical observatory, for which purpose the expedition took along a 120-mm telescope. On the basis of the very comprehensive measurements made, the south-eastern plateau of the Pirkuli, a tableland 1,435 metres above sea level, in the Caucasus was recommended. The terrain is 22 km from the village of Shemakha and is about 160 km to the west of Baku. On 13 January 1960, the Presidium of the Academy decided to build the astronomical observatory on the

Pirkuli and 160 people, including 55 scientists and 25 technicians, now work there.

One of the first tasks of Shemakha Observatory was the investigation of the Sun. For the monitoring of the chromosphere and photosphere of the Sun, the astronomers there have been able to use the "AFR-2" telescope since 1957. The "photospheric tube" has an aperture of 200 mm while that of the "chromospheric tube" is 130 mm. The monitoring of the chromosphere is carried out with the aid of polarization interference filters of 0.05 mm band-width.

A horizontal "AZU-5" telescope was set up in the Department for Solar Research in 1963 and its primary and secondary heliostat mirrors have a diameter of 440 mm. This instrument is employed for the spectroscopic and photometric investigation of the Sun. It is used with a grating spectrograph, mainly for the investigation of the magnetic fields of sunspots and of the brief, periodic fluctuations in brightness of the flocculi. The Solar Department specializes in the investigation of sun flares and their relationship with the magnetic field of the Sun and in the study of the fine structure of the solar atmosphere.

The observational resources of the Department for Stellar Physics were significantly improved in 1966 with the commissioning of a 2-metre reflector built by VEB Carl Zeiss, Jena. This instrument can be used in the prime, Cassegrain and coudé foci. The prime focus is employed for direct photography, with a grating spectrograph for spectroscopic observations. A prism spectrograph, a grating spectrograph and a star photometer are available for the Cassegrain focus. The coudé spectrograph is equipped with three cameras having focal lengths between 350 and 1,400 mm. With these numerous ancillary instruments, the telescope can be used for a wide variety of tasks.

The Department of Stellar Physics is concerned with problems of the T Tauri stars—irregular young variable stars—, of close double stars, magnetic stars and hot stars with an expanded outer shell. In addition

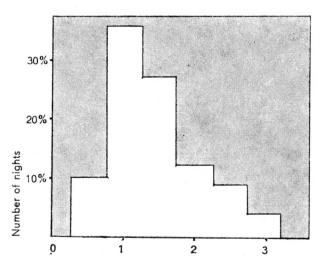

Diameter of the star image in seconds of arc due to air turbulence

14   Investigations of the atmosphere and image quality were carried out on Pirkuli Mountain in 1953. The illustration shows that on most nights the diameter of the seeing disc was less than 1.5 seconds of arc. The maximum distribution of diameters is at one second of arc.

to the 2-metre telescope, there is also a photometric "AZT-8" telescope available for research tasks in the field of stellar physics. The latter instrument is a 70-cm paraboloid reflector with a focal length of 2,820 mm and can be used in two different Cassegrain variations.

The "AST-452" meniscus telescope with a Maksutov system was taken into service in 1965 and is used for spectrographic and photometric observations, especially in the ultraviolet spectral region.

The 2-metre telescope is not however for the exclusive use of the Department of Stellar Physics at Shemakha and since 1969 spectral observations of the planets with a high dispersion have been carried out with this instrument. Numerous spectra with a dispersion of 0.6 nm per millimetre have been acquired of Mars, Jupiter, Saturn and Uranus. Among other things, the fine structure of the methane distribution on Saturn has been investigated and it has been demonstrated that the atmosphere of Saturn cannot contain any ammonia or, if it does, then it must be in the crystalline form. The investigation of the radial distribution of hydrogen on Jupiter revealed that at a distance of 0.75 radii from the centre the intensity of the hydrogen line was 60 % less than at the centre.

A special department in Shemakha Astrophysical Observatory is concerned with astronomical history.

The Observatory of the Azerbaidzhan Academy of Sciences is used by many visiting astronomers from other Soviet and foreign observatories. The acquisition of a 90-cm Schmidt telescope is planned for the near future to supplement the equipment of the Observatory.

● The Astrophysical Observatory in the Crimea was founded at the beginning of the 20th century and works closely with the Main Astronomical Observatory at Pulkovo near Leningrad. In the last century, significant contributions were made at Pulkovo to astrometry and the investigation of double stars in particular but the observing conditions at Leningrad were never satisfactory. The famous "White Nights" do not allow much astronomical observation in summer and the sky over Leningrad is often overcast in autumn and winter. Consequently, it was suggested at a very early date that an observing station should be established in the south of Russia. In 1906, A. P. Ganski, an astronomer from Pulkovo, made a journey to the Crimea and, in the little town of Koshka near Simeiz, found a private observatory which was used primarily in the interests of popular science. It was immediately agreed that the observatory in the Crimea should function as an outstation of Pulkovo Observatory. The Observation Department was officially established near Simeiz in 1912. In the same year, a 1-metre telescope was ordered for the new station but it was 1925 before the new instrument could be installed, the first photographs being taken with it on 28 May 1926.

The 1-metre telescope was used by G. A. Shain to make comprehensive observations which subsequently formed the basis for an important catalogue of radial velocities. Numerous spectroscopic double stars were discovered with the instrument and their orbits determined. From these observational programmes, it was soon apparent that the site was not suitable for astronomical observations. It was at 350 metres above sea level. In the south, the terrain slopes downward towards the town of Koshka while in the north it rises to a height of 600 metres. Cold currents of air from the mountains meet warm and moist masses of air from the Black Sea, resulting in the forming of clouds and serious air turbulence.

The equipment at the Observatory suffered major damage during the Second World War. The reconstruction of the Simeiz Observatory with numerous observational instruments took place after 1946 at a more suitable site. The most important instruments and the date of their commissioning are as follows:

1949 40-cm double astrograph with a focal length of 160 cm, a field of 10° by 10° and a plate format of 30×30 cm. An objective prism is also available for the photographic instrument.
1950 Coronagraph for the monitoring of the Sun
1952 1.2-metre reflector for spectroscopic work
1953 50-cm Maksutov reflector
1954 Solar tower
1961 2.6-metre telescope
1967 22-metre radio telescope for work in the millimetre range. The radio telescope is not located at the Observatory but in a bay on the Black Sea not far from Simeiz.

At the Crimean Observatory of the Academy of Sciences of the USSR, as it is now known, there are four departments:

1. Solar and Planetary Physics,
2. Stellar and Nebular Physics,
3. Radio Astronomy,
4. Experimental Astrophysics.

The Crimean Observatory is one of the most important solar observing stations in the USSR. The solar tower is the principal facility. A coronagraph of the Lyot type is operated with narrow-band filters which also permit the observation of the helium line at 1,083 nm, i. e., in the nearer part of the infrared. The optical observations are carried out in parallel with the monitoring of the Sun in the radio-wave region at 1.5 metres.

The principal instruments of the Stellar Physics Department are the 1.2-metre and the 2.6-metre telescopes. The spectroscopic observations with the 1.2-metre reflector are used in particular for the investigation of young hot stars, the nature of the stellar atmosphere and the chemical composition of the stars of the

various spectral classes. The 2.6-metre telescope is employed for the exploration of distant galaxies, especially active galaxies and their nuclei, and for the investigation of quasars.

The purpose of the Department for Experimental Astrophysics is to carry out laboratory investigations and to pursue research into the evolution of spectra and the excitation of atoms in hot gases under different physical conditions. Another important area of activity of this department is extraterrestrial observational astrophysics. It is here that equipment is built for observation outside the Earth's atmosphere. Lunokhod 2 was equipped with a photometer for measuring the brightness of the sky on the Moon which was made in the Crimean Observatory. Interestingly enough, these measurements showed that both the night sky and the day sky on the Moon are significantly brighter than expected.

The Department for Experimental Astrophysics is also involved in observations of space vehicles. Their actual orbits are compared with the calculated data and corrections to their orbits worked out from this information. With the electron-optical systems and using television, the Mars 6 probe, for example, was identified even at a distance of 465,000 km when it was only of a brightness of the 17th magnitude.

● Few observatories are so firmly linked with the name of their founder as Sonneberg Observatory. Today, it is part of the Central Institute for Astrophysics of the Academy of Sciences of the GDR and it has been greatly extended since 1945. However, the field of activity and the enthusiasm of the staff for astronomy still clearly reflect the profound influence of C. Hoffmeister (1892–1968).

C. Hoffmeister, the son of a businessman of Sonneberg, was determined right from his early youth to enrich astronomy with serious contributions, mainly on the basis of his own observations and he allowed none of the unpropitious circumstances of his life to obstruct this aim. At the age of sixteen, he began with the systematic monitoring of meteor phenomena. In the course of his life, he devoted many nights to this aspect of astronomy, enriching it with important theoretical findings and finally, at the age of 55 years, summarizing it in a standard scientific work. It was with particular energy that he devoted himself to the investigation of variable stars. His name was first mentioned in 1914 in connection with the discovery of such an object but by the end of his life he had discovered and studied several thousand variable stars through untiring observations at the telescope and intensive evaluation of observational material.

Although C. Hoffmeister worked for a time as an assistant at the Remeis Observatory in Bamberg and at the University Observatory in Jena where the quality of his work was demonstrably acknowledged, his wish was always to establish more independent and, above all, more favourable observing conditions at Sonneberg. It was a hard struggle to achieve this but it demonstrated his ability and will-power when the town of Sonneberg and the region of Thuringia were ultimately persuaded to make modest resources available for the erection of a "Thuringian Observatory". At the time it was officially opened, in 1925, on the Erbisbühl, a southern spur of the Thuringian Forest of 640 metres in height, it was the highest observatory of Germany.

The programme on which the present-day importance of this research institute is mainly based—the systematic search for variable stars—dates back to the early days of Sonneberg Observatory. C. Hoffmeister always took the view that a star with a constant brightness was really an exception and that a photometric check of adequate sensitivity of any star should show fluctuations. Important information about stars as astrophysical objects was therefore to be expected from the precise study of such a natural phenomenon and modern developments have fully confirmed this hypothesis. To begin with, however, this required the acquisition of observation material on a comprehensive scale in the form of chronological brightness records—light-curves—of as many variable stars as possible. This material had to be particularly extensive since the variability of these stars is not due to identical physical processes: these bodies belong to a whole series of groups. The latter differ in their outer characteristics, for instance, and this is evident from observations. For this reason the collection of basic material for the study of the variable stars was the declared aim of the "Monitoring of the Sky" programme and the "Area Plan" of Sonneberg Observatory.

The "Area Plan" was a project for the regular photographing of selected areas of the sky with the maximum range possible. Through careful comparisons in a comparator of plates taken at different times, it should then be possible to identify all the variable stars down to the brightness limit. Such completeness encourages, for example, statistical evaluation according to the relative frequency of the individual types which, in turn, has major cosmogonic consequences. A 40-cm astrograph was used for the "Area Plan" photographs. This type of instrument is rather suitable for a programme of this kind both on account of the large field of view supplied by the lens system and because of the quality of the image definition. After the war, the programme of work was successfully continued with a new 50-cm Schmidt telescope and two new astrographs.

The "Monitoring of the Sky" programme, which was likewise started in the early days of the Observatory and is still being vigorously pursued, is of a similar character to the "Area Plan". However, it is not restricted to the regular photographing of individual areas since in this case several small cameras are used on clear nights to photograph the whole of the accessible sky. With this programme, there is no question of reaching the stars of very low apparent brightness. Nevertheless, the monitoring plates have about the 15th magnitude as their limit and up to this there are ten million stars in the Sonneberg sky.

The inestimable value of the 150,000 photographic plates taken in the course of the "Monitoring of the Sky" and "Area Plan" projects consists in the period of decades for which the behaviour of each one of the objects photographed can be traced back or which can be used even for identifying stars with very slow changes in their brightness. There has occasionally been disparagement of astronomy carried out with modest instruments but this kind of observation is necessary and its worth is always proven when it enables objects which have become interesting to be traced back for decades. Outstanding examples of this are the quasars, cosmic sources of X-ray radiation, and special variable stars which have suddenly become very important for theorists since they change their variability, become constantly brighter or so on.

The situation today is not the same as when C. Hoffmeister began the work on the "Area Plan" and the "Monitoring of the Sky" project. Thanks to the efforts of astronomers, including those at Sonneberg who alone discovered almost a third of all known variable stars, there is now a very comprehensive amount of basic material at the disposal of science. It is now a question of the interpretation of the objects concerned in the sense of their physical state. This is why the activities of Sonneberg Observatory, apart from the work on the "Area Plan" and the "Monitoring of the Sky" project, are now centred on systematic observations and theoretical investigations in connection with variable stars and their interaction with the interstellar matter around them. This work includes the taking of photoelectric measurements in various wavelength areas, such as the infrared, and the recording of the spectra of special stars. For the photometric work, suitable equipment had first to be developed, this being done in the Observatory's own workshops and electronics laboratory.

Research on variable stars and the search for unknown bodies of this kind call for reliable "book-keeping" on already known objects. This is something which has been carried out very meticulously at Sonneberg Observatory. The corresponding index was once incorporated in the bibliography *History and Literature of the Variable Stars* and now, via the Stellar Data Centre in Strasbourg, it can be called off by any astronomer in the world.

In addition to the telescopes erected on the Erbisbühl, to which a 70-cm Cassegrain telescope for photometric observations was added only a few years ago, the eleven astronomers at Sonneberg Observatory can also use the 2-metre universal telescope of the Karl Schwarzschild Observatory, Tautenburg.

● Nowadays, a modern giant telescope can detect about a thousand million stars and stellar systems. Although it is true that astronomical research is restricted to the investigation of a suitable selection of examples from this profusion, there is a constant accumulation of new data material on a colossal scale. Night after night, the observatories of the world, with constantly improving methods, are taking measurements which enable the most diverse conclusions to be drawn about the objects studied. Thus, in general, the same results can be utilized for the solution of different kinds of problems. This is why it has always been customary for astronomers to publish and exchange on a worldwide scale not only scientific results but also the actual data acquired. In view of the extent of the material already available or constantly being published for the first time in the form of lists and more or less comprehensive catalogues, this kind of collaboration is no longer sufficient. The individual scientific worker is finding it increasingly more difficult to keep in touch with the available material, even in his own specialized field, and duplication of work is often unavoidable. In addition, working with data collections is a time-consuming business. It has to be accepted, for example, that a star can be mentioned in the literature on the subject under as many as twenty different designations in the form of catalogue-numbers. This identification problem alone justified the effort, at a central institution, of establishing order in the place of chaos. The entire material had first to be put in the appropriate form for a computer with a big memory, which involved a lot of work, but it was then rapidly accessible at any time.

It was on the basis of such considerations as these and at the suggestion of some well-known European astronomers, that a data centre was founded by the French "Institut National d'Astronomie et Géophysique" in 1972 and set up at the observatory of the Louis Pasteur University in Strasbourg. Of course this "Centre de Données Stellaires" has its own staff of technical and scientific workers but it relies on close collaboration with specialists at other astronomical establishments. This results, for example, from its function which includes, among other things, the critical evaluation of observational data. This means that a definitive figure has to be determined from various data which differ from each other by reason of unavoidable errors of measurement but refer to the same quantity for one and the same star. This cannot be done by a purely mathematical procedure but requires a more intensive examination of the individual results and thus experience. It is for this purpose that photoelectric measurements of star brightnesses and colour indices are processed at the observatories in Lausanne and Geneva. The task of the Astronomical Calculating Institute at Heidelberg is the cataloguing of star positions and proper motions, the Sonneberg Observatory of the Central Institute for Astrophysics of the GDR collects data on variable stars and Marseilles Observatory processes the data concerning measured radial velocities of stars. Finally, the Paris Observatory maintains a bibliography of the stars in which, classified according to celestial bodies, are listed those literature references which contain astrophysical data on about 70,000 stars. Rapid access of this kind to fundamental publications is an important pre-condition for all scientific work. The numbers handled at the Stellar Data Centre are incredibly large: astronomical data is available for 1.1 million stars, there is information on the spectrum, i. e., spectral type, radial velocity or rotation, of 480,000 bodies and photometric data on 170,000 stars can be provided. In general, access to this data is via a catalogue which solves the problem of identification mentioned above. For 450,000 stars, a designation according to any astronomical nomenclature can be used and the computer then indicates the data for precisely that star in question.

The central memory for the Data Centre is a giant computer installation at Meudon Observatory near Paris. A 400-km long data line links the Stellar Data

Centre with this computer. Since various French observatories likewise possess direct access to this computer, the entire collection is immediately available to them. All other astronomical establishments can request data and then receive it in the form of print-outs, tapes, punched cards, micro-films or even star charts which indicate the positions of the objects in different colours within a network of coordinates.

The Stellar Data Centre is a very recent foundation which, in the future, will certainly also collect data from other branches of astronomy and thus will become increasingly more important in a central sense.

● Sunspot is the name which the scientific and technical workers at Sacramento Peak Observatory gave to their residential community in the Sacramento Mountains in the southern part of New Mexico. Together with their families, they put up with the altitude and the isolation—it is 50 km from the nearest town of any size, Alamogordo—in order to be able to use the excellent seeing conditions of the site for the investigation of the Sun.

The founding of the Observatory dates back to the time when, about 30 years ago, it was realized by the US Air Force just how much influence the Sun has on our atmosphere. Solar activity phenomena can lead, by way of physical processes in the ionosphere, to a complete collapse of radio wave propagation and to disturbances in the Earth's magnetic field. Both factors represent a hazard for aerial and space navigation. Thus, initially, the principal task of the Observatory was the processing of observational and theoretical programmes for forecasting disturbances caused by the Sun. In 1976, the Air Force transferred the administration of the Observatory to the National Science Foundation which, in turn, handed it over to the Association of Universities for Research in Astronomy (AURA) but its original function is still maintained. From the astronomical viewpoint, this means the acquisition of more and more information about the "laboratory" of the Sun and its interaction with matter in interplanetary Space and with the planets themselves. Work is concentrated on the investigation of the corona and the study of the most delicate details on the visible surface of the Sun. The corona is an envelope, several solar radii in extent, of highly rarified gases which are heated to a temperature of one million Kelvin by surge phenomena from the visible surface. It is not possible to observe this extremely faint manifestation at the same time as the bright disc of the Sun. However, it is visible when a total eclipse of the Sun occurs since the disc of the Sun is then covered by the Moon. For the investigation of the corona, eclipses of the Sun are too few and far between, of course, and so the astronomer arranges an artificial eclipse with the aid of a small stop in the light-path of a "coronagraph". Nevertheless, if the solar corona is to be visible, it is also essential that the atmosphere above the point of observation be free from haze and dust particles causing scattered light. In this respect, Sacramento Peak Observatory is in an ideal location. This is also the reason why several coronagraphs are in operation here to monitor the corona, to take photometric exposures of it in the light of different spectral lines and to measure the polarization of its radiation. This type of instrument was developed by the French astronomer B. Lyot in 1930. The coronagraph with which the first observations were made on Sacramento Peak in 1949 and which is still in use was the second of its kind in the Western Hemisphere.

The great solar tower characterizes the silhouette of Sacramento Peak. Dazzlingly white, it soars to a height of 40 metres above ground-level and is thus unaffected by the layer of turbulence just above the ground which results from solar thermal radiation and which would be a serious disturbance to observations. So that the tower itself does not become a source of troublesome air turbulence, it is painted with suitable reflecting paint and there is a special cooling system in its concrete walls. From the tip of the tower, the rays from the Sun are passed perpendicularly downwards to a depth of about 100 metres via two movable flat mirrors more than a metre in diameter and they then arrive at the principal mirror of the telescope which gives an image about 50 cm in diameter in the solar laboratory. Details with a size of a quarter of a second of arc can be identified and investigated with special equipment.

15 a/b The cross-section through the tower telescope gives an indication that the technical problems are to be found in the details and not in the basically simple design. The enlarged detail shows the heliostat, a combination of two plane mirrors, the first of which in the light-path (right) follows the varying altitude of the Sun on its daily journey and, with the entire head, turns around an axis which runs vertically through the second mirror. This motion takes account of the variable azimuth of the Sun.

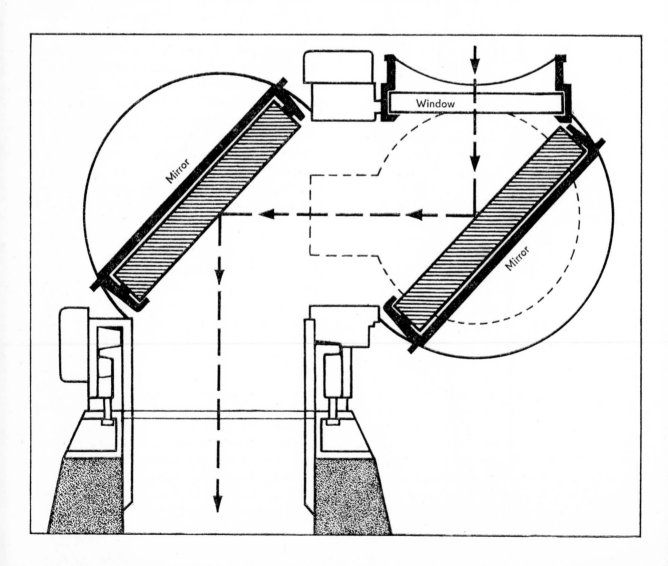

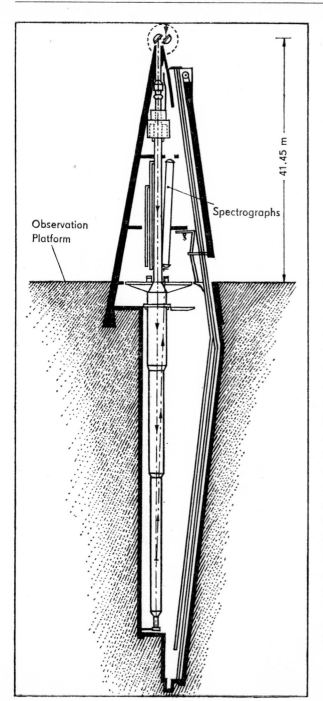

Observation
Platform

Spectrographs

41.45 m

This angle corresponds to 1/7500 of the apparent diameter of the Sun and illustrates what accuracy is necessary for pointing to the object; it must be maintained despite the daily movement of the Sun across the sky. This is carried out by a photoelectric tracking system which controls the entrance-mirror. The light-beam from the tip of the tower to the detection equipment (photographic plate or photoelectric photometer) passes through a vacuum and the entire telescope—a steel tube of 100 metres long, up to three metres in diameter and weighing 250 tons—can be rotated around its perpendicular longitudinal axis with a high degree of accuracy. This rotational movement is also necessary because, with the chosen mounting of the two plane mirrors on the tip of the tower, the image of the Sun turns with the passage of time. The maintenance of the adjustment of the solar image for long measuring-periods can only be assured by the simultaneous rotation of all the measuring equipment connected to the telescope tube and of the 120-sq. metre platform. All movements are of course computer-controlled, like almost all of the other operations of this complex telescope. The arrangement of the entire light-path in a vacuum is necessary if image disturbances by internal currents and layers of air of different temperatures are to be avoided. The consequence of this requirement of the astronomers was a whole series of technical problems resulting from the size of the volume to be evacuated, the rotational movements of the telescope in its entirety and of the entrance-mirrors and from the necessary control operations at the equipment also working in a vacuum. The two pumps need about six hours to reach the operating level, equivalent to the air pressure at a height of 55 km above ground-level.

At the present time, Sacramento Peak Observatory scientists are considering how the equipment can also be used for the night observation of cosmic bodies.

Institute of Astrophysics and Atmospheric Physics of the Estonian Academy of Sciences, Tartu, USSR

Astronomical research in the Estonian SSR is concentrated at the Institute for Astrophysics and Atmospheric Physics, an establishment of the Academy of Sciences of this republic. The main astronomical work is carried out at the W. Struve Astrophysical Observatory, located at about 20 km to the south-west of Tartu. The areas of work of the Institute include the physics of individual cosmic objects, problems of the structure of our stellar system, the investigation of entire groups of such systems and questions concerning the structure of Space. Investigations of a purely theoretical nature and practical, photometric and spectroscopic procedures complement each other in this.

The principal observational instrument is a 1.5-metre reflector which was taken into service in 1975. It was supplied by the LOMO Optical Works in Leningrad, well-known for its construction of the giant Soviet telescopes. The instrument is equipped with the appropriate spectrographs and photoelectric photometers for various wavelength areas. Some of this equipment was developed at the Institute. For observations over a long period of measurement, an automatic tracking and guidance system is used, this monitoring the exact movement of the telescope. It is envisaged that a specially programmed computer will soon take over other control functions.

The building of the old observatory is situated right in the modern city of Tartu and is part of the present Institute. It was erected in 1811 as the observatory of Dorpat University which had been founded a few years before. After St. Petersburg, it was the second oldest university of Russia. The establishment was headed by F. G. W. Struve from 1818 to 1839, his name being associated above all with the founding of the astronomy of the double stars. He was also probably the first actually to measure the distance of a fixed star, this being Vega in the constellation Lyra. For this, he utilized the annual parallax shift of the closer star against the background of faint and far distant objects, appearing as the counterpart of the annual motion of the Earth

around the Sun. Generations of astronomers had endeavoured to find the proof of this effect which had already been identified by Copernicus as a necessary consequence of and evidence for his heliocentric concept. However, the slight change in position remained unnoticed until the 19th century since it was less than the margin of error present in the instruments until then available. F. G. W. Struve carried out his observations between November 1835 and August 1839. At about the same time, F. W. Bessel in Königsberg, as it was then called, was likewise working on the determination of a fixed star parallax. Admittedly, he had planned his measurements at an earlier date but was only able to actually take them between August 1837 and October 1838. Such parallel developments are not seldom in the history of science and they show that the time is ripe—in this case especially in the technical respect—for such a discovery. It even happened that a third astronomer, G. A. Henderson in the Southern Hemisphere, announced a parallax determination immediately after F. W. Bessel and F. G. W. Struve.

The success of F. G W. Struve and the importance of Dorpat Observatory were based on the instruments and equipment available at this research establishment. In connection with the purchase of a meridian circle from Reichenbach's Mechanical Workshop, F. G. W. Struve travelled to Munich and, on this occasion, also visited the Optical Institute of J. von Fraunhofer and J. von Utzschneider. He summarized the impression he received there in the following words: "I saw an achromatic telescope of a size never attempted by the English and this was fabricated with great care and perfection in all its parts." The significance of this opinion is apparent when it is realized that the Englishman J. Dollond had invented lenses which were largely free from chromatic aberration and that England in general at this time held a leading position in the manufacture of lenses. But J. von Fraunhofer, through the systematic use of special glasses, improved techniques in the production of lenses, precise testing-methods and through the emphasis on mechanical precision, achieved a major advance in the construction of astronomical instruments. The refractor ordered by Struve in Munich arrived at Dorpat in 1824 and, with a lens diameter of 24 cm, it was the largest of the day. Apart from this, however, this instrument was regarded as a wonder for many years on account of its technical innovations. It was the first large telescope with an equatorial mounting and was equipped with a clockwork mechanism to compensate for the daily rotation of the sky. That the instrument aroused the enthusiasm not only of the experts but also of laymen is apparent from the fact that it was exhibited in a church in Munich before being despatched to Dorpat. F. W. Bessel also used one of Fraunhofer's instruments for his work.

With the installation of the Reichenbach meridian circle and the Fraunhofer refractor, Dorpat was the best-equipped observatory of its time. F. G. W. Struve made full use of the opportunities there and, within two years of the erection of the refractor, published a catalogue of more than 3,000 double-star systems, comprising the results of a systematic search for such objects. This catalogue was supplemented in 1837 by accurate micrometric measurements of the relative positions of the components of these double stars, this representing the first material for the subsequent derivation of the movement which the members of such a double-star system perform around each other.

Today, after the necessary restoration, the Dorpat refractor is now on show in the East Hall of the old observatory as the prize exhibit of the Astronomical Museum, a department of the Tartu Municipal Museum, where it can be seen by the public.

● The Karl Schwarzschild Observatory was founded on 19 October 1960 and is part of the Central Institute for Astrophysics of the Academy of Sciences of the GDR. Apart from the Karl Schwarzschild Observatory, the observatories at Babelsberg, Potsdam and Sonneberg also belong to the Central Institute for Astrophysics.

The Observatory is situated at 350 metres above sea level in the vicinity of Tautenburg, a village which dates back 750 years, and is about 15 km from the university city of Jena, the home of the Zeiss works. The proximity to VEB Carl Zeiss and to the Friedrich Schiller University is an advantage for the work of the Observatory but, on the other hand, the presence of a city in the neighbourhood causes a brightening of the night sky and thus problems for astronomical observations. For some time, a Government order has been in force providing a protective zone for the Observatory and guaranteeing not only that observing conditions will not deteriorate further but even that there will be an improvement.

The Karl Schwarzschild Observatory at Tautenburg is equipped with the first 2-metre reflector to have been supplied by VEB Carl Zeiss, Jena. The development and construction of this telescope was the product of a long Jena tradition in this field. The production programme of the manufacturer in the area of astronomical instruments thus extended from school and amateur telescopes and various reflectors of the 1-metre class to the 2-metre universal reflecting telescope. Further instruments with a 2-metre free aperture have been supplied in the meantime by VEB Carl Zeiss, Jena, to observatories in Czechoslovakia, the Soviet Union and, more recently, Bulgaria.

The Tautenburg telescope can be used in three optical variations. Since one of these variations is a Schmidt camera, the mirror must have a spherical surface. The corrector plate is 140 cm in diameter and has a free aperture of 134 cm, thus making this telescope the largest Schmidt camera in the world. The field of view is 3.5° by 3.5° and the plate format 24 cm × 24 cm.

The Schmidt camera has a focal length of four metres. This special optical system was pioneered by Bernhard Schmidt, an Estonian optical worker, who combined a spherical mirror with a correction lens for the first time at Hamburg in 1930 and thus obtained a large field of view with an undistorted image. The corrector plate is located in front of the mirror at a distance equivalent to the radius of curvature. The focal length of the system is half as great as the radius of curvature so that the image is produced exactly midway between the mirror and the corrector plate. As the focal plane of a Schmidt camera is not flat but a spherically curved area, the photographic plates must be curved in their plate holder when an exposure is made.

The Tautenburg Schmidt telescope is equipped with a second corrector plate. This is prismatically ground and therefore does not depict the stars in the usual form of spots but instead as short spectra. This permits an effective search for specific types of stars as the basic material for certain investigations. The equipping of an astronomical telescope with an objective prism is nothing new but this is the first time that the maker has combined the optical functions of a corrector plate and a prism in a single component. This avoids light-losses and thus makes this telescope even more powerful. The instrument is used as a Schmidt camera only for the two weeks around the new moon in each month since the marked brightening of the sky from moonlight otherwise prevents long exposure times with such a high-speed system. The telescope is then used more effectively in one of its other two optical modes, the Cassegrain or the coudé arrangement.

For the Cassegrain focus, which is accessible at the two transom ends of the fork mounting, there are two spectrographs available. One of these is used in combination with multi-stage image intensifiers which produce a substantial gain in range. The focal length of the Cassegrain system is 21 metres.

In the coudé variation with its focal length of 91 metres, the light-beam is passed via several auxiliary mirrors through a fork transom into the polar axis and then into the basement of the dome building. This is the permanent location of a large, temperature-stabilized spectrograph with which a reciprocal linear dispersion of up to 0.2 nm per millimetre can be obtained.

The primary function of Tautenburg Observatory is to lay the observational basis for the current research work at the Central Institute for Astrophysics. For the Tautenburg astronomers, this means involvement in the investigations of extragalactic stellar systems and in the programmes for the investigation of stars displaying magnetic fields. In general, the Schmidt camera is used for the former whereas the study of the magnetic stars is based on numerous individual spectrograms, each of which has to be acquired by hours of exposure time on the spectroscopic equipment.

The archives of the Karl Schwarzschild Observatory now contain more than 5,000 Schmidt exposures and almost 3,000 photographic plates with spectra. The scientific staff of the Observatory also use the facilities of the telescope and the material already available for their own astronomical projects, of course. However, they are devoting very careful attention to increasing the performance of the telescope and its radiation detectors and to improving the methods for the acquisition of information from the actual observation material. For the evaluation of astronomical photographs and spectrograms, Tautenburg has at its disposal an iris photometer, a spectrum photometer, blink and spectrum comparators, a coordinate measuring instrument and—for the further processing of data—a computer unit.

A well-equipped photographic laboratory, installed in the dome building, is of great importance for the work of the Observatory. This is not only able to process the photographic yield of a night's observations but can also investigate the suitability of various photographic emulsions for certain purposes and is involved in the development of new processing techniques for photographic materials. Here, too, the aim is to improve the efficiency of astronomical observations.

● The Astronomical Observatory in Tokyo is one of the 14 institutions of Tokyo University and was founded in 1878. However, it remained at its original site on the university campus for only ten years and was moved to Azabu, an urban district of Tokyo, in 1888. The expansion of the city and the rapid deterioration in seeing conditions resulting from this forced another move in 1920, this being to Mitaka Village, now known as Mitaka City and a part of the Tokyo conurbation. The headquarters of the Astronomical Observatory are indeed still there but the individual observational facilities are now distributed throughout the entire country.

The Observatory of Tokyo University, now employing a staff of more than 200 persons, including 50 astronomers, has always been concerned with practically all the important branches of astronomy. This is immediately apparent from the instrumentation of Mitaka Observatory. Here there are a 65-cm refractor with an equatorial mounting, a 20-cm meridian circle and a 20-cm photographic zenith telescope. These instruments are used for precise astrometrical observations which are still carried out at Mitaka since Mitaka is the headquarters of the national time service. For the exact determination of time, the passages of stars through the meridian are observed with meticulous accuracy and the data compared with the indications of precise atomic clocks.

A 48-cm solar telescope was also used at Mitaka for many years but is no longer in service now. The outstation of Norikura at an altitude of 2,876 metres above sea level was opened in 1951 specifically for optical solar observations. In a 5-metre dome, there is a 10-cm coronagraph of the Lyot type and a photoelectric K-coronameter on a single mounting there. A 25-cm coronagraph with coudé focus is housed in a 12-metre dome. In combination with a Littrow spectrograph, it is possible to achieve a reciprocal linear dispersion of 0.05 nm per millimetre, i. e., the two known yellow lines of sodium which are spaced at 0.00000006 cm in the spectrum, are more than 1 cm apart in this spectrograph after the splitting up of the light.

For observations in the millimetre range in particular, use is made of a parabolic reflector 6 metres in diameter at Mitaka. However, the presence of a city like Tokyo in the vicinity causes not only a brightening of the night sky from electric lighting, which seriously disturbs optical observations, but also a high noise level in the radio-frequency region. It was therefore only logical that a new station, specially for radio astronomy

16 The Astronomical Observatory of Tokyo has establishments throughout the country. Its headquarters are at Mitaka and its five observing stations are in districts with good conditions for astronomical observations.

observations, should be built 150 km to the west of Tokyo in November 1970. One of the principal instruments is the radio interferometer whose east-west arm consists of nine 6-metre paraboloidal antennas and two 8-metre paraboloidal antennas. The north-south arm of the interferometer is made up of six 6-metre reflectors. All reflectors have an equatorial mounting. "Radio photographs" of the Sun can be obtained with this instrument.

The multi-element interferometer at Nobeyama, which has twelve 1.2-metre reflectors and works on a wavelength of 1.8 cm, is mainly used for the observation of the distribution of radio waves across the disc of the Sun.

At Okayama, 600 km to the south-west of Tokyo, an observing station was opened in 1960 which, from the very beginning, was planned as a facility for use by astronomers from all Japanese universities and institutes. The principal instrument of this national observation centre is a 1.88-metre reflector with Newtonian, Cassegrain and coudé foci. Other instruments at Okayama include a 91-cm Cassegrain telescope and a 65-cm solar telescope. Since scientists from all institutes can carry out observations at Okayama, the ancillary equipment for the telescopes must allow a variety of programmes to be carried out. For the Cassegrain focus of the 1.88-metre telescope, there are two two-prism spectrographs, a grating spectrograph with image intensifier and a grating and an échelle spectrograph for the coudé focus. Linear dispersions of between 0.13 and 12 nm per millimetre can be attained in the coudé focus. The observation programmes at the telescopes at Okayama range from investigations of the solar system to extragalactic research.

The observing station at Dodaira, 60 km away from the headquarters of the Observatory, was completed in 1962. Here, too, there is a 91-cm telescope for direct photography at the prime focus and spectroscopic and photometric work at the Cassegrain focus. However, the special feature of this station is the Baker-Nunn camera which was previously installed at Mitaka: Dodaira is thus a point in the international satellite-observation network. The Astronomical Observatory of the University of Tokyo also works closely with the Institute for Space Travel and Aeronautical Science in Japan. For this, Dodaira is equipped with a laser satellite tracker which fires light pulses with one joule of energy at the satellite. The reflected signals are picked up with a 50-cm Cassegrain reflector. This method permits very accurate distance determinations of the satellites. A 5-joule laser tracker and a 1.5-metre metal reflector are also under construction at Dodaira as detectors for reflected signals.

The fifth outstation of Mitaka started operations only in the autumn of 1974. It is at a height of 1,120 metres above sea level in the Kiso Mountains, 200 km to the west of Tokyo. There are excellent atmospheric conditions for astronomical observations here. In the Kiso Mountains, there is a Schmidt telescope with a free aperture of 105 cm and it has proved to be a great stimulus for galactic research at Tokyo Observatory.

Since Tokyo Observatory is an establishment of the University, the staff there also have teaching duties at the University and are responsible for the training of future astronomers in Japan.

● The town of Torun does not have an astronomical tradition of its own but is mentioned time and again in an historically significant connection. It was here that Nicolaus Copernicus was born on 19 February 1473 and lived in the midst of well-to-do merchant families until he was 18. He then began his studies at Cracow and various universities in Italy. In the course of his further life, which was devoted to the interests of his government posts and to the development of his heliocentric concept of the world, he did not return to his home town for any length of time. Despite this, Torun still merits the fame for having produced the man whose work brought about a turning-point in science.

Modern astronomy in Torun began after the Second World War. When the Nicolaus Copernicus University was opened in the town, where the great astronomer was born, two chairs were established for the study of astronomy. These led to the founding of the present-day Institute with its observatory and departments of astrophysics and stellar astronomy, radio astronomy and celestial mechanics. Within the Institute, there is also an astrophysical laboratory which is affiliated to the Polish Academy of Sciences.

The beginning of practical astronomy at the new University of Torun after the Second World War was only made possible by the loan of a telescope. Harvard College Observatory of Cambridge, Mass., provided a 20-cm refractor with two objective prisms. This was the famous astrograph which, at the turn of the century, had been used to acquire the photographs on which the spectral classification of the 225,000 stars of the *Henry Draper Catalog* is based. The site chosen for the instrument, which was the nucleus of the now greatly expanded observatory, was the village of Piwnice, 12 km to the north-west of Torun in a flat part of the countryside.

The Draper Telescope caused the research interests of the Institute to be dominated by spectroscopic investigations and, in the area of optical astronomy, this is still the case today. At the beginning of the 1960's, a larger telescope was taken into service. The 90-cm mirror of this instrument can be used as the principal component in two different optical systems. The first of these is the Schmidt camera, for which the entrance-aperture is reduced to 60 cm. At Piwnice, when the instrument is used in this version, it is mostly in conjunction with an objective prism, i.e., the photographic plate forms an image of the spectra of all the stars which can be reached within the field of view. Of the investigations carried out at the Institute on this basis, those in particular that have won international recognition are concerned with the assignment of objects to certain groups within our stellar system—the populations—and with classification of objective prism spectra according to more than one parameter. Originally, only the surface temperature of the stars, for the most part, was derived from such spectra with their relatively low resolution. At Torun, however, it was a question of using this material to acquire information about pressure, chemical composition and movements in the stellar atmospheres and thus to make the fullest use of the information content of this method which is very effective in respect of observation.

In a second optical version, a slit spectrograph can be used in the Nasmyth focus of the 90-cm telescope. With the spectra of high resolution thus acquired, investigations carried out with objective prisms can be supplemented and the astronomers can tackle problems which call for slit spectrograms from the very start. In this latter area of activity, astrophysical work has been carried out at the Observatory concerning the outer layers of certain types of stars. Considerable research has also been devoted to the members of our planetary system, especially the spectra of Venus and Mars but with attention also being paid to comets in the course of these investigations. In the past, most of these spectrograms came from foreign telescopes which were put at the disposal of Polish astronomers but since 1974 the Institute has possessed its own high-power spectrograph. This was developed and built at the Dominion Astrophysical Observatory at Victoria, Canada, on the basis of decades of experience. It was handed over to the University of Torun on the occasion of the ceremonies commemorating the fifth centenary of the birth of Nicolaus Copernicus.

On the terrain of the Observatory at Piwnice and in the direct vicinity of the domes of the optical telescopes, there are the antennas of the large radio astronomy department. Research in this sector began in 1958 with a tiltable parabolic antenna of 12 metres in diameter. To begin with, the changing radiation of the entire Sun was monitored on a wavelength of 2.3 metres. So that smaller centres of activity on the Sun could be isolated, improved radio interferometers have been brought into service over the years. These give an increased spatial resolution by coupling the signals received at several separate antennas. In the simplest case, used in 1960, the two receiver installations were arranged in the form of cylindrical paraboloids with the dimensions of 8 $\times$ 4 metres which were set up in the east-west direction with a separation of 24 metres. Since the resolving power of the installations increases with the distance between the antennas, it was only logical that these should be moved further and further apart in the course of time. The ultimate result of this will be an interferometer of five large and fully movable paraboloids which will be set up on an east-west base of 3 km in length. Sophisticated receiver technology and on-line control by computer will bring about a further upswing in the quality of research. The first antenna of the system with a reflector diameter of 15 metres was taken into service in 1977. The investigations of the radio astronomy department at Torun are still concentrated on the Sun but in the meantime other important results have been achieved in the measurement of the radio emission of the satellites of Jupiter, in the study of interstellar plasma with the aid of pulsar signals and in the mapping of cosmic sources of radio waves.

Since the radiation of celestial bodies with a wavelength of more than 20 metres cannot be observed from the Earth on account of the limits set by terrestrial at-

mosphere, radio astronomers at Torun University developed a radio spectrograph for use in artificial terrestrial satellites. This was launched in 1973, two months after the fifth centenary of the birth of Nicolaus Copernicus, in the "Intercosmos Copernicus 500" satellite by the Soviet Union. In collaboration with Soviet scientists, measurements were taken of solar radio waves on long wavelengths. The reduction and interpretation of the data was carried out under the direction of Polish radio astronomers.

● The investigation of the bodies of the planetary system has always been one of the major tasks of astronomy. From Antiquity until the beginning of modern times, it was only a question of the investigation of the motions of these celestial bodies. With the formulation of the heliocentric view of the world by Copernicus and the discovery of the laws of motion by Kepler, the planets were put on the same level as the Earth. The first minor planets (asteroids) were discovered at the beginning of the 19th century and it was not long after this that evidence was shown of the relationship between comets and meteorites. However, with the investigation of the bodies of the planetary system, astronomy became increasingly involved in the examination of their surface structures, the composition of their atmospheres and finally the question of their internal structures and their origin.

With the launching of artificial satellites and especially space-probes, many believed that the exploration of the bodies of the planetary system would no longer be Earth-based but would be done exclusively with the new space-probe technology. This is not at all the case, as is clearly demonstrated by the founding of the Lunar and Planetary Laboratory and above all by the instruments with which it is equipped. This is expressed very vividly by the words of G. P. Kuiper on whose initiative this research and teaching institute at the University of

88 The 2.6-metre telescope has been operating in the Crimea since 1961. At that time, it was the largest telescope in Europe and was the first giant telescope to be built in the USSR. The experience gained with this instrument was of eminent importance for the other telescopes subsequently designed and built in the Soviet Union.

Page 158:

89 This general view of the Crimean Observatory shows only some of the instruments located there. The favourable position of the Observatory on a large plateau in a heavily wooded region is clearly evident.

90 The 22-metre radio telescope has an altazimuth mounting. It is situated in a bay of the Black Sea. This instrument has a fully lined reflector surface and is suitable for observations in the millimetre range. The instrument is operated jointly with the optical telescopes so that observations of celestial bodies can be carried out simultaneously in different spectral ranges.

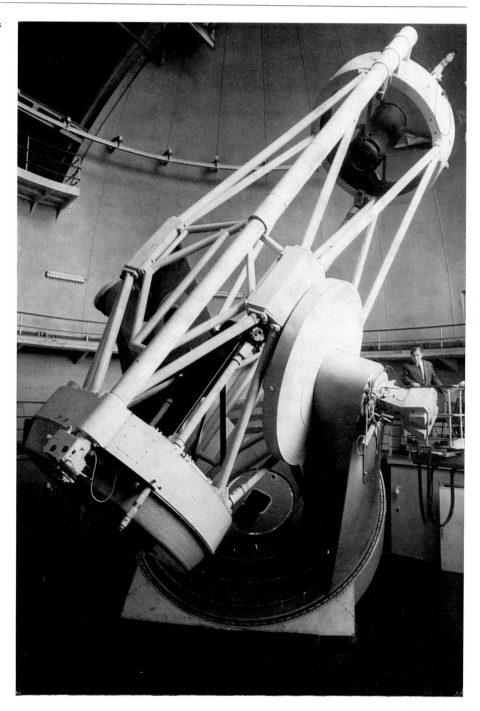

91   Partial view of Sonneberg
Observatory. It is administered
by the Central Institute for
Astrophysics of the Academy
of Sciences of the GDR.

92   C. Hoffmeister (1892–
1968), the founder and for many
years the director of Sonneberg
Observatory.

Pages 160/161:
93   As part of the Sonneberg monitoring programme
for variable stars, a total of
14 cameras on two mountings
"produce" up to 200 photographs of the sky on clear
nights.

94   With a photometer developed at Sonneberg Observatory
for a 60-cm telescope, the radiation of stars in the visible and
infrared spectral regions can be
measured with two special
photoelectric detectors.

95 The University Obser-
vatory is situated in the old
quarter of Strasbourg, in a kind
of park. The dome building in
Renaissance style was erected
about 1880. It accommodates
a 49-cm refractor, the objective of which was supplied by
Merz of Munich and the mounting by the well-known firm of
Repsold of Hamburg.

96 The "entrance" to the
vacuum solar tower of Sacramento Peak Observatory.

Page 164:
97 Measuring platform in the
solar tower of Sacramento Peak
Observatory. Parts of the
evacuated telescope tube can
be seen on the left of the picture.

98   The old principal building of Dorpat Observatory has
survived in its original form
but is now surrounded by the
city of Tartu. In the tower,
there is a fairly small telescope
which is still used for astronomical observations. In the left
wing of the building, an opening
can be seen which passes
through the wall and the roof
and provides the meridian
circle with a field of view run-
ning parallel to the meridian
line.

Pages 166/167:
99   Once a miracle of precision engineering, the Dorpat
refractor is now a much-admired
exhibit in the astronomical museum. A massive stand carried
the parallactic mounting with
the wooden tube of 4 metres in
length. F. G. W. Struve justified the use of wood by a
reference to the good temperature behaviour of this material.
The two balls of lead at the
eyepiece end prevent the tube
from sagging when the telescope

is in various positions. Looking
through the stand, a view can
be had of the weights used to
drive the clockwork mechanism
when observations were in
progress so that the entire
instrument followed the daily
rotation of the sky around the
inclined polar axis aligned with
the celestial pole, the stars
sighted remaining for fairly long
periods in the field of view.

100   The new 1.5-metre telescope from the Leningrad
Optical Works presents an at-
tractive, functional appearance.
Measuring facilities are provided at the prime, Cassegrain
and coudé foci. The picture
shows a photoelectric photometer at the Cassegrain focus,
access to which from the end of
the telescope is provided via
the pierced principal mirror.

101   The 2-metre telescope of
the Central Institute for Astrophysics is housed in a 20-metre
dome at the Karl Schwarzschild
Observatory, Tautenburg. The
white paint of the dome building and an insulating layer in
the 180-ton dome ensure that
sunlight does not cause a heat
build-up inside the dome as
compared with night-time
temperatures.

Pages 169/170:
102   The large Cassegrain
spectrograph can be seen on
the fork transom nearest to the
camera. Dispersions of 17.5 to
3.5 nm per millimetre can be
obtained with it. The total
moving weight of the telescope
amounts to 65 tons. The instrument is carried on a thin film
of oil so that a motor of only
55 W is required for the tracking movement. The telescope
can be operated from the prin-

cipal control console in the
dome, from a mobile observation platform in the dome and
also from the coudé rooms
situated in the basement of the
building. With the aid of coordinate display instruments,
the telescope can be directed at
any point in the sky even from
the coudé rooms. The movement of the 5-metre slit in the
dome is automatically syn-
chronized with that of the
telescope.

103   This copy of a Schmidt
exposure of the Orion nebula
taken with the Tautenburg
telescope was produced with
the aid of a special photographic contrast-control technique. This enables the extended structures of the nebula to
be seen very clearly and details
can even be identified in the
heavily exposed central parts
of the nebula.

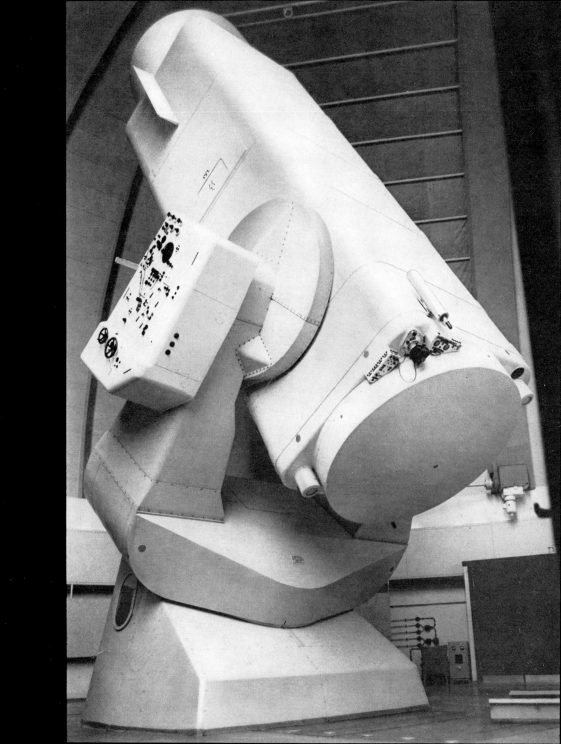

104   The 20-cm meridian circle at Mitaka is used for the observation of star passages through the meridian. Since sidereal time is determined by the culmination of the vernal equinox and the vernal equinox is simultaneously the zero point of the right ascension coordinates, the observations of the passages of stars can be used for the determination of time.

Pages 172/173:
105   The 1.88-metre reflector of Okayama can be used as a Newtonian telescope with a focal length of 9.2 metres, as a Cassegrain system with a focal length of 33.8 metres and in the coudé focus with a focal length of 54.5 metres.

106   The 1.5-metre metal reflector of Dodaira is used as a receiver for the laser pulses emitted from Dodaira and reflected back by artificial satellites of the Earth or by the Moon. The distance to the "reflector" can be determined from the time taken by the pulses. Since a very high degree of accuracy can be achieved, it is possible to demonstrate even slight changes in the orbits of satellites.

107   Norikura Observatory at a height of 2,876 metres is linked with the nearest community, Takayama, by a paved road, 40 km long. It is specially equipped for the observation of the Sun.

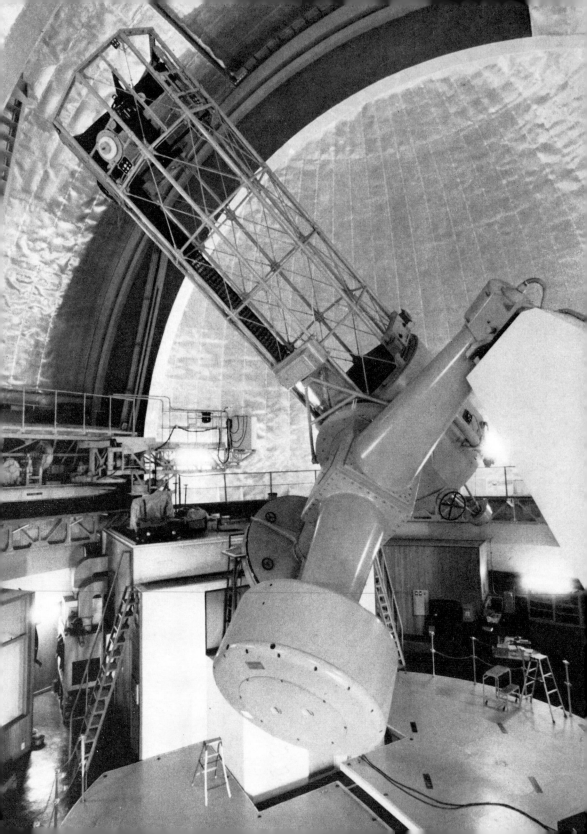

108   Nicolaus Copernicus
(1473–1543)

109   Copernicus was born in
this house in present-day Torun
on 19 February 1473.

110   The picture shows the
90-cm telescope at the Observatory of Nicolaus Copernicus
University. The guide telescope
used during observations for
checking is attached to the main
instrument. It is a refractor
with a 20-cm aperture. The
astronomer is looking through
the finder which, with its wide
field of view, facilitates pointing at the object to be observed.

111   The Lunar and Planetary
Laboratory comprises numerous
technical departments. One of
the most important of these is
the optical workshop where the
1.55-metre mirror was ground.
The large grinding-machine
can be seen in the picture.
On the grinding-table there lies
the blank of the 1.55-metre
mirror. The photograph was
taken shortly before the grinding operation was started.

Page 177:
112   The 1.55-metre telescope
has an English frame mounting.
The check ring for the various
attachments can be seen at the
end of the tube. These attachments include a planetary
camera, a photo-polarimeter
and an interferometer.

Page 178:

113 With an equatorial diameter almost ten times that of
the Earth, Saturn is the second
largest planet of the solar
system. The ring system of this
planet extends for more than
278,000 km but is not more
than 15 km thick. This exceptionally thin annular disc is
formed from dust particles
which revolve around the central body as "mini-moons".
Until recently, this ring was
considered unique in the planetary system but similar structures have been identified in
the meantime around Jupiter
and Uranus.

114 This Viking 1 picture
(1976) of the surface of Mars
does not appear strange at all.
Stoney, reddish areas such as
this can also be found on the
Earth. The relatively sharply
defined horizon is a consequence of the extremely thin
atmosphere of Mars. The pressure on the surface of Mars is
only about 10 mb, i. e., about
one hundredth of the pressure
on the Earth's surface. The
reason for the low density of
the Martian atmosphere is the
low mass of the planet which is
only about 0.107 of the mass
of the Earth.

115   The 4-metre telescope
has a horseshoe mounting which
has proved especially suitable
for large and heavy instruments.
The 4-metre mirror, which is
60 cm thick, weighs 15 tons
by itself.

Page 180:
116   The dome of the 4-metre
telescope stands on the highest
point of Kitt Peak. The total
height of the building is
60 metres and it is 35 metres
in diameter.

117   The McMath Solar Telescope is used not only during
the day for observations of
the Sun but also at night-time
for investigations of the planets
and stars.

118 The cylindrical part of the dome building of the 1.8-metre telescope is 11 metres in height. The building and the dome are 22 metres in diameter. The slit in the dome is 5 metres wide. The building, the dome and the slit-cover are double-walled to ensure through good ventilation that a heat build-up does not occur in the interior of the dome during the day. The building of the 1.8-metre instrument also accommodates a mechanical workshop, a very important part of an observatory for repair and maintenance work and for the construction of new ancillary equipment for the telescope.

Page 182:
119 The 1.8-metre telescope has an open, skeleton tube. Although this gives the instrument a light appearance, the total moving weight equals 45 tons. The hour-wheel of the telescope can clearly be seen. Behind the principal reflector, there is the Cassegrain spectrograph and, in the background, the time-keeping installation for controlling the tracking movement of the telescope.

120 The 1.2-metre telescope can be used as a coudé system. The plane mirror, which directs the beam of light through the hollow axis and into the coudé rooms, can be detected at the tip of the polar axis.

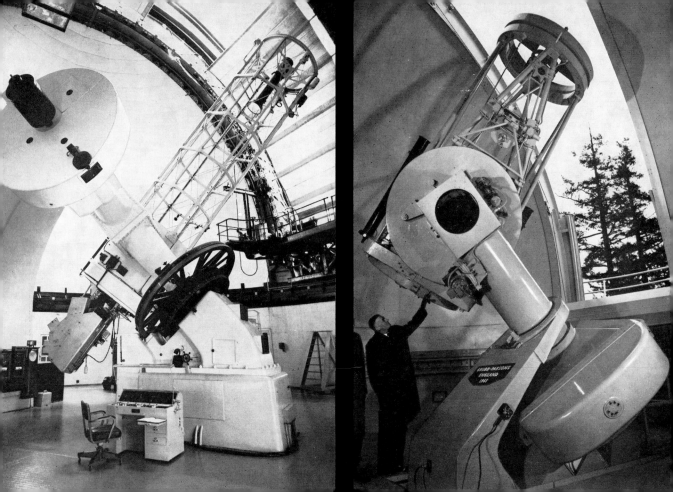

121   The slit-cover of the 40-
metre dome of the 6-metre
telescope is not moved to the
side in the customary manner
but swung over the dome.

Page 184:

122   With the aid of a large
crane, the main components of
the telescope were lowered
through the slit in the dome for
the final assembly. The crane
has been kept for any subsequent assembly work which
may be necessary.

Arizona was established in 1960: "The problems and methods of the exploration of the Moon and the planets are very similar to those of geophysics. The geophysicist must concern himself with the application of physics, chemistry, dynamics, hydrology, aerodynamics, photo-chemistry and other basic sciences in order to interpret the geophysical phenomena observed. A similarly broad range of the physical sciences must be used for the study of the planets and their atmospheres. However, on account of the distance of these bodies, the methods must be used indirectly, basing on phenomena which are observed from a distance. Under this aspect, the technique is analogous to and often identical with that of stellar astronomy. The use of telescopes—Earth-based, in balloons and aircraft and in space-vehicles—is very similar, but the problems involved are very different. All in all, planetary astronomy is concerned with cold matter, gas or solid particles in molecular or crystalline form. While our problems are similar to those of geophysics, the technique is that of stellar astronomy."

The astronomical instruments for Earth-based observations are at three outstations. 54 km away from the headquarters of the Lunar and Planetary Laboratory on the campus of Arizona University, there is the observatory at Catalina at an altitude of 2,515 metres above sea level. Three telescopes are stationed here, the principal instrument being a 155-cm reflector for high-resolution observations. This has been in operation since 7 October 1965. It was with this telescope that the photographs for the *Consolidated Lunar Atlas* with a resolution of 0.15 seconds of arc were acquired. G. P. Kuiper published a *Photographic Lunar Atlas* in 1960 and this was complemented by the *Orthographic Atlas of the Moon* in 1961 and by the *Rectified Lunar Atlas* in 1963. The 1960 atlas is a collection of the best photographs of the Moon from Mount Wilson, Lick, McDonald, Yerkes and Pic du Midi observatories. The first supplement is a combination of the most suitable photographs of the Moon with an improved coordinate system for the Moon. In the improved lunar atlas of 1963, special importance is attached to the crater distribution, structure lines and ray systems on the Moon. This atlas was designed specifically for the interpretation of photographs of the Moon taken from space-vehicles. In the *Consolidated Lunar Atlas* of 1967, particular attention has been paid to photographs with maximum resolution of specific formations on the Moon.

The Schmidt telescope of Catalina Observatory has a free aperture of 42 cm and is used especially for the observation of comets.

About 500 metres from the dome of the 155-cm telescope, there is a 1-metre telescope on a small elevation and this is employed for photometric observations. From this it is evident that the Lunar and Planetary Laboratory is concerned not only with planetary astronomy but also with stellar photometry.

A unique infrared observatory has been built some 65 km away from Tucson. The station is located on Mount Lemmon at a height of 2,800 metres. The site is characterized by exceptionally low humidity and is thus particularly suitable for work in the infrared spectral region. The observation of planetary bodies in the infrared region is particularly important since the maximum radiation at the surface temperature of many members of the planetary system is in this wavelength range. At a temperature of $0°$ C, the maximum radiation is at 0.01 mm and at $400°$ C at 0.004 mm.

All the observation instruments at Mount Lemmon Observatory are designed for work in the infrared. The largest instrument is a 150-cm reflector. It is known as the "150-cm C" and is the third in a series of metal reflectors developed by the Lunar and Planetary Laboratory. The prototype, the "150-cm A", is installed at the San Pedro Martir Station in California. On the site of Mount Lemmon Observatory, there are also instruments belonging to other research establishments but they are likewise used for infrared observations.

The Universities of Minnesota and San Diego operate a 150-cm telescope on Mount Lemmon. The primary purpose of infrared observations on Mount

Lemmon is the study of the bodies in the planetary system but attention is also paid to special stellar objects, such as Chi Cygni. In the visible spectral region, the brightness of this star is about four decimal powers less than the bright star Vega while on the photographic plate it is six decimal powers less. At 1.3 µm wavelength, there is little difference between the two stars but at 2.2 µm Chi Cygni is five times brighter than Vega.

If it is largely infrared observations which have to be made, an altitude higher than two or three thousand metres is desirable to avoid the troublesome influence of the Earth's atmosphere, especially water-vapour absorptions. This is why the Lunar and Planetary Laboratory lost little time, after its foundation, in beginning work on "balloon astronomy". Telescopes with a diameter ranging between 30 cm and one metre were taken to a height of 36 km with the aid of large balloons. Observations of the planets were carried out with an automatically guided 30-cm telescope and with a liquid helium cooled bolometer in the range of 5 to 1,000 µm, i. e., from the close infrared to the radio-frequency area. This programme led to the discovery that the planets Jupiter and Saturn radiate three times as much energy as expected from the absorbed solar radiation.

Photographs from a height of 24 km were also taken of the Earth. 98 % of the terrestrial atmosphere is then below the "observer". Observations from this altitude are very important for the understanding of noctilucent clouds and of the photochemical production of oxygen in the higher strata of the atmosphere. An institution which displays such determination in the exploration of the planetary system will also utilize the most advanced aid at its disposal for this—space travel. The scientists of the Lunar and Planetary Laboratory were involved in the development of the programme and in the evaluation of all lunar space missions from Ranger to Apollo and of the flights to Mercury, Venus and Jupiter. The artificial Moon craters, as caused by Rangers VII, VIII and IX for example, were investigated in great detail by the staff of the Laboratory. Since precise details

were known of the kinetic energy of the objects striking the Moon, it was possible to draw very accurate conclusions in respect of the formation of natural impact craters on the Moon.

The programme of work of this research establishment covers practically all the problems encountered in the exploration of the planetary system. This is why not only astronomers but also geophysicists and chemists work at the Lunar and Planetary Laboratory.

Another point which must be mentioned is that work of very great value has been done by this institution regarding the selection of good observational sites. For example, a very intensive test programme was carried out by scientists from the Laboratory in the Mauna Kea area of Hawaii at an altitude of 4,155 metres and it was partly due to this work that a decision was taken to build modern observatories on this site.

● Kitt Peak National Observatory (KPNO) is about 90 km from Tucson, Arizona. It possesses the largest concentration of observational instruments for the exploration of the Universe in the Northern Hemisphere. Kitt Peak National Observatory is sponsored by the National Science Foundation and the Association of Universities for Research in Astronomy. The plan to establish a National Observatory was drawn up about 25 years ago and the search for a suitable site began in 1955. The criteria for this site were: a large number of clear nights, a minimum of air turbulence, remoteness from bright light sources but not too far from a university if possible, uniform vegetation over a fairly large area and as high an altitude above sea level as possible. In all and in the course of three years, 150 potential sites in the western states of the USA were examined before the decision was taken in 1958 to build the centre for optical astronomical observations on the 2,064-metre high Kitt Peak in the Papago Reservation in Arizona. To begin with, the Indian inhabitants of the area opposed the building project on Kitt Peak. They feared that it would be a source of disturbance to them and especially to the spirits of their gods since they believed that a rocket-launching installation was to be constructed there. The astronomers took some of the elders of the tribe to the Steward Observatory of the University of Tucson, showed them their work and instruments and allowed them to look through the telescopes. Relations improved and a treaty was signed which will remain in force as long as scientific work is carried out on Kitt Peak. However, the caves around the summit of the mountain are out of bounds to the astronomers since the Indians believe that Ea-Ee-Toy, the god of the Papagos, is sometimes to be found there in the form of some creature or other Ea- Ee-Toy normally lives on Baboquivari Peak, about 20 km to the south of Kitt Peak, and the Indians consider this latter mountain to be the centre of the Universe.

Within a few years, Kitt Peak developed into an astronomical observation centre with, at the present time, twelve telescopes at its disposal. This "host" of telescopes is headed by the 4-metre Mayall Reflector which, after the 6-metre reflector of the Soviet Union and the 5-metre reflector on Palomar Mountain, is one of the biggest in the world, and by the largest solar telescope with its aperture of 1.5 metres.

The headquarters of the Institute in Tucson were built at the same time as the observatory on Kitt Peak and it is here that the administrative centre, the mechanical and optical workshops, the electronics laboratory, a computer centre and of course a library are located.

A committee headed by N. U. Mayall met for the first time on 23 September 1961 to discuss the urgent construction of a 4-metre telescope on Kitt Peak. Definite specifications were drawn up by the end of that same month:

1. The distortion-free image of an area of 0.75° to 1° in diameter was to be available at the prime focus.
2. An aperture ratio of 1:2.8 was to be provided at the prime focus.

3. The necessary optical correction systems for the prime focus exposures were planned for use in the ultraviolet spectral region.
4. The Cassegrain focus was to be in the form of an optical Ritchey-Chrétien system.
5. In the interests of effective observation even on moonlit nights, provision was made for a coudé system with a horizontal spectrograph.
6. The use of quartz for the principal mirror and all the secondary mirrors was proposed.
7. Intensive studies on the local seeing led to the decision to site the instrument about 45 metres above ground-level.
8. Provision was to be made for the simple and rapid changing of systems.

The telescope was completed by 1973 at a cost of ten million dollars. The first photographic plate was exposed in the prime focus in March 1973. The dedication ceremony took place on the 19th and 20th of June in Tucson and on Kitt Peak and the instrument was named the "Mayall Telescope". In Tucson, the dedication was coordinated with a public symposium—the "Copernicus 500".

In addition to the 4-metre telescope, the world's biggest solar telescope is also in operation on Kitt Peak. The 2-metre heliostat mirror is situated 31 metres above ground-level on a tower. This passes the light from the Sun through an inclined shaft to the paraboloid 1.5-metre mirror which is located about 30 metres below ground-level. The part of the shaft above ground-level is enclosed in a water-cooled jacket. There is a distance of 153 metres between the heliostat mirror and the paraboloid mirror. The paraboloid mirror passes the light from the Sun along the same shaft but at a slight angle in relation to the incoming beam and again towards ground-level. A 1.2-metre mirror there then directs the beam perpendicularly into the observation room which is located below ground-level.

The solar telescope has a focal length of 92 metres and produces a solar image of 76 cm in diameter. For the analysis of the beam, the solar telescope is equipped with a vacuum spectrograph which is a duplicate of the Michigan spectrograph.

250 visiting astronomers come to Kitt Peak every year from all parts of the USA and from all over the world. However, an institution such as this attracts not only the experts but also members of the general public who have a keen interest in astronomy. The Observatory is open to visitors every day who want to learn more about the work carried out there. Account of this requirement was taken even at the design stage of the buildings and domes of the great telescopes and galleries have been provided for visitors. A tour through Kitt Peak Observatory begins with an introductory film-show in which an impression is also given of the interesting work of grinding and polishing the 4-metre mirror. The terrain of the Observatory also contains instruments of the University of Tucson which cannot be visited during the day but evening tours are organized once a week instead when it is possible to look through the telescopes. There is also a museum on the site providing information about the development of astronomy, its tasks and the results achieved.

Kitt Peak provides excellent conditions for astronomical observations but a serious problem was encountered during the construction of the Observatory—the question of a water supply. This was solved by asphalting a large area of the terrain for the collection of rain-water which is then stored in large tanks.

It is not possible to cover the entire sky from Kitt Peak or the other observatories in the Northern Hemisphere; large parts of the Southern Sky remain inaccessible. This is why Kitt Peak Observatory was given a counterpart in the Southern Hemisphere. With the assistance of the National Science Foundation and the National University of Chile, work began in 1963 on the construction of the Inter-American Observatory on Cerro Tololo in Chile. This mountain is 2,200 metres high and is situated 480 km to the north of the capital, Santiago. Ideal observing conditions are found here.

There is a total of eight instruments on Cerro Tololo at present. The biggest of these is a 4-metre telescope, a duplicate of the 4-metre telescope at Kitt Peak. The mirror, which was likewise ground in the optical workshop of Kitt Peak Observatory, is not made of quartz but of Cervit, a material with an even lower coefficient of thermal expansion than quartz. A library, the most important workshops, guest houses and catering facilities are also available on Cerro Tololo. Every year, many astronomers, especially from the USA, come here to acquire observational material of the Southern Sky.

● The first photographic spectrum was taken with the 1.8-metre telescope in the Canadian astrophysical observatory in Victoria on the 8th of May 1918. This was the beginning of the Dominion Astrophysical Observatory, Victoria, a research institute now known throughout the world. Even at this time, it already had a forerunner in Canada, this being the Dominion Observatory, Ottawa, which was founded in 1905.

On the basis of practical and economic considerations, the Canadian Federal Government began to support astronomical research in the last decades of the 19th century. A transcontinental rail-link had to be built which, in particular, required a great deal of map-plotting work and accurate geographical positional determinations on the basis of astronomical observations. The importance of astronomy for national development was thus recognized and it was decided to establish an observatory at the seat of government in Ottawa. This Institute is the Dominion Observatory which was completed in 1905.

For its tasks which were primarily concerned with positional astronomy, it was equipped with a 45-cm telescope but it was also possible to use this for spectroscopic work. During the directorate of J. S. Plaskett, outstanding results were achieved in studies of the motions of the stars. It was from its work on orbits of double stars in particular that the Dominion Observatory achieved worldwide recognition. J. S. Plaskett and his colleagues also investigated the physical characteristics of stars. The new questions which followed from this work called for a larger telescope, especially for the spectroscopic investigation of the fainter and more distant stars.

In 1913, the decision was taken to build a new giant reflector and it was originally planned that the new telescope should also be erected at Ottawa. However, to avoid any restriction at all of the performance of the instrument, the astronomers recommended that a site with better meteorological conditions should be sought for the telescope. The choice fell on a district about 20 km from Victoria where an elevation of 250 metres in height was selected for the Dominion Astrophysical Observatory and was given the name of Observatory Hill. The 1.8-metre telescope which was commissioned there in 1918 was the largest in the world at that time but for 40 years it remained the only telescope of the Dominion Astrophysical Observatory. The 1.8-metre telescope is arranged as a Cassegrain system with a focal length of 36 metres, i. e., it has a focal ratio of 1:20. The Cassegrain focus is behind the principal mirror which is pierced in the centre to allow the beam of light to pass through. Visual, photographic and spectroscopic observations are possible at the Cassegrain focus. The first spectrograph for the 1.8-metre telescope was designed by J. S. Plaskett, the founder of the Observatory and it was a prism spectrograph. The prisms have now been replaced by gratings which are more efficient.

After the Second World War, the Observatory received another instrument with a 1.2-metre mirror and of the same basic construction as the 1.8-metre instrument. However, this second instrument has, in addition to the Cassegrain focus, a coudé focus, permitting the installation of stationary, air-conditioned spectrographs of high resolving power.

The spectrographs of both telescopes are equipped with image slicers and electronic image intensifiers. As a result, the spectrographs of this Canadian observatory are some of the most powerful in the world although the telescopes are by no means the largest.

An important phase in the development of the Dominion Astrophysical Observatory was the building of a large optical workshop capable of grinding mirrors up to 4 metres in diameter. One of the first "products" of this shop was a new mirror for the 1.8-metre telescope. The original mirror was made of crown glass which was subject to dimensional deformation in the presence of temperature differences between the front and the back of the mirror and this had a detrimental effect on image quality. A new mirror of Cervit was made for this instrument. This is a glass-ceramic material with a coefficient of thermal expansion which is even less than that of quartz.

To achieve better resolution, to reach faint objects and thus penetrate further into Space, Canada and France completed a joint project with the University of Hawaii and built a 4-metre telescope on Mauna Kea. The mirror for this instrument was likewise ground in the optical shop of the Dominion Astrophysical Observatory. However, the 4-metre telescope does not belong to the Observatory but is at the disposal of all Canadian astronomers.

19 scientific and 14 technical workers are pursuing a variety of astronomical tasks at the Dominion Astrophysical Observatory. Even in the past, investigations were made of galactic structure problems with special attention being paid to young stars and novae and work has also been done on questions of double stars and the radial motion of stars. More recently, the astronomers of this Institute have turned their attention to variable stars, particularly in connection with X-ray sources, Beta Cephei stars and the examination of the frequency distribution of chemical elements. Mention must also be made of the important work done on instrument problems associated with the construction of the 4-metre telescope on Mauna Kea.

● For the time being, the construction of ever larger optical instruments has come to an end with the commissioning of the 6-metre telescope of the Academy of Sciences of the USSR. The plan for the construction of this giant telescope was put forward by the Astronomical Council of the USSR and dates from 1960. From previous experience in the design and construction of giant telescopes, the total weight of the 6-metre telescope was estimated as almost 1,000 tons. The stability and uniformity of the movement of such a great weight imposes very exacting requirements on the design of the mounting. The first plan envisaged an equatorial mounting for the 6-metre telescope, as used up to then for all giant telescopes. The great advantage of the equatorial mounting is that the telescope only needs to be moved around the polar axis, which is parallel to the Earth's axis, in order to follow the apparent daily movement of the stars from east to west. The disadvantage is that the polar axis is at an angle in relation to the zenith. The angle between the polar axis and the horizontal plane corresponds to the geographical latitude of the site. From this inclined position, asymmetries result on account of the varying position in relation to the direction of gravity in the movements. With the great weight of the 6-metre telescope, the flexure problems arising from the asymmetries were of such a magnitude that the original plan for an equatorial mounting was abandoned.

The altazimuth type of mounting is symmetrical. One axis of rotation is perpendicular to the horizontal plane while the other is parallel to the horizontal plane. When the telescope is moved, the weight will thus always act on the bearings in the vertical direction and the direction and amplitude of any flexing will thus remain constant. However, for tracking, the instrument must be moved simultaneously and at different speeds around the two axes. This is handled by a computer. It continuously converts the equatorial coordinates of the object under observation into horizontal coordinates, takes account of elevation-dependent refraction and also controls the rotation of the exposure area which, with an altazimuth mounting, rotates once in 24 hours. The symmetry of the altazimuth mounting also makes it lighter than the equatorial mounting and less expensive. It is estimated that an equatorial mounting for the 6-metre telescope would have doubled the cost.

Although the total weight was reduced by the altazimuth mounting, the 6-metre telescope still weighs 800 tons. The axis of rotation perpendicular to the horizontal plane and carrying the entire weight is supported on oil under pressure, i. e., a great ring on this axis is carried on six oil "cushions". The entire instrument floats, as it were, on a film of oil 0.1 mm thick. Due to this oil-bearing system and the complete symmetry of the loadings involved, only a small angular momentum is needed to turn the telescope with its weight of 800 tons in the various horizontal directions. The actual telescope is carried in a fork located on the perpendicular axis of rotation. A force of only about 3 N is required to incline the telescope at the various angles of elevation.

The mirror for the 6-metre telescope was ground and polished in the Leningrad Optical Works and is made from Pyrex glass with a very low coefficient of expansion. Nevertheless, with a block of glass 42 tons in weight, six metres in diameter and 65 cm thick, fluctuations in temperature lead to distortions which have a detrimental effect on image quality. For this reason, a proposal that a new mirror should be made from Sital, a glass-ceramic material, is now under consideration. When the blank for the 6-metre mirror was cast in the 1960's, it was not then possible to make such large blocks of Sital. The coefficient of expansion of Sital is even less than that of quartz.

At the back of the mirror, 60 holes have been bored in four concentric circles for 60 stress-relieving systems. As a result, the changes in shape caused by the mirror's own weight are less than 1/16 of a wavelength, i. e., less than 0.00003 mm.

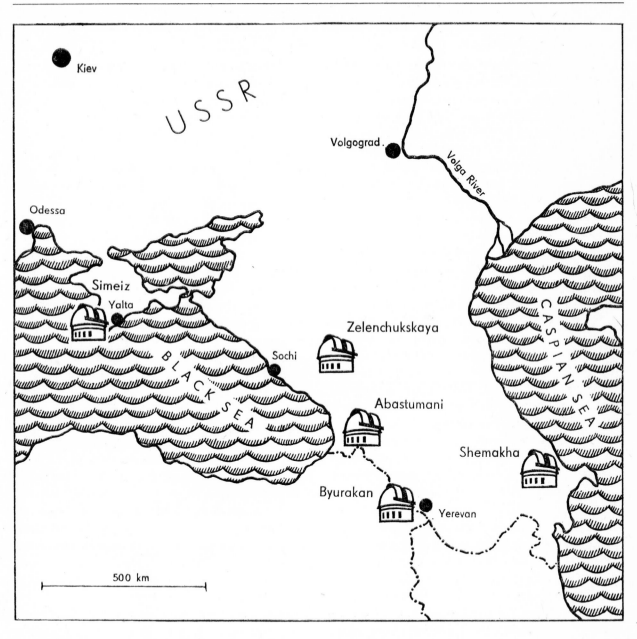

17 Zelenchukskaya is situated in the region between the Black Sea and the Caspian Sea where there are also the observatories of Byurakan, Abastumani and Shemakha. Zelenchukskaya is 2,070 metres above sea level. Experience so far has shown that there are about 70 photometric and 170 spectroscopic nights per year here.

Two optical systems can be used with the 6-metre mirror, which has a focal length of 24 metres. Individual objects can be photometrically and spectroscopically observed there. When a supplementary optical system is positioned in the light-path immediately in front of the focal point, a field of view of about 0.3° in diameter is obtained for direct photography. All the ancillary instruments for the prime focus are located with the observer in a spacious prime focus cabin 1.8 m in diameter and more than 2 m long in the middle of the telescope. The cabin obscures only 9 % of the surface of the principal mirror.

A Nasmyth system can also be used with the 6-metre mirror. The hyperbolic secondary mirror required for this is 76 cm in diameter and the focal length of this system is 186 metres. A plane mirror in the horizontal axis of rotation is used to bend the light-path through the tips of the fork transoms and to the observation platforms. One platform is utilized for small ancillary instruments, such as photometres, spectrographs, image intensifiers and so on, whereas a large spectrograph is installed in the other fork transom, this permitting a dispersion of 0.1 nm per millimetre.

The 6-metre telescope has a guide telescope with an aperture of 70 cm and a focal length of 12 metres with which, for exposures taking a long time, the tracking of the telescope can be visually checked and, in the event of deviations, manually corrected. Automatic photoelectric guidance of the telescope is also possible, however. In this case, the computer takes over the monitoring function and gives instructions for corrections. In this manner, the computer checks its coordinate calculations itself, so to speak.

In the few years which have passed since the telescope was taken into service, experience has shown that the 800-ton telescope can be guided to an accuracy of 0.2 seconds of arc both manually and with the automatic photoelectric method.

So far, with direct photography in the blue spectral region, it has even been possible to reach objects of the

18 This design sketch shows details of the 6-metre telescope.
CM—centre of mass.
PM—primary mirror,
SM—secondary mirror,

OF—oil-film,
NF—Nasmyth focus,
OP—observation platform,
SS—small spectrograph,
LS—large spectrograph.

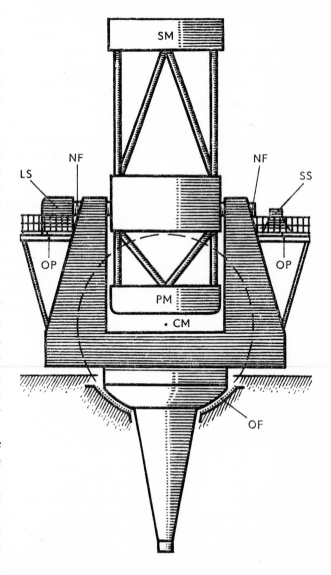

24th magnitude after an exposure time of 30 minutes while in the red spectral region and after 60 minutes even stars of the 23.5th magnitude have been photographed.

Many Soviet and foreign astronomers have already paid a visit to the telescope since, in the opinion of many experts, it represents a significant stage in astronomical instrument design in two respects. Firstly, it will remain the world's largest optical telescope for a very long time to come, to say the least, and secondly, the success of the altazimuth mounting constitutes a new approach to the construction of giant telescopes.

Within the next few years, Zelenchukskaya will develop into one of the world's most important research centres and this will be due not only to the 6-metre telescope but also to the modern RATAN radio telescope (RATAN = Radio Astronomy Telescope of the Academy of Sciences). The RATAN radio telescope consists of 900 parabolic plates, forming a circle of 600 metres in diameter. Each plate can be individually adjusted to guarantee an exact circle. The telescope can be used with all 900 plates as a detector system or each quarter of the plates can operate as a self-contained unit. Interesting new research results can be expected from the combined use of the 6-metre and RATAN telescopes.

Concerning the Dutch telescope: The Dutch telescope is the first invention of its kind which in the hands of a Galilei has yielded extraordinary benefits and has disclosed to us things in the heavenly bodies, of which the ancients had either absolutely no idea or only an imperfect conception.

"Wöchentliche Hallische Anzeige",
Monday, 14th July 1760.

● There have been many advances in the development of the sciences resulting from technological progress. The leap forward in quality in the field of astronomy, which was a consequence of the use of space travel for research, is also a part of this. The man-made missiles orbiting the Earth or traversing the planetary system became experimental proofs of the celestial mechanical laws in Space and served as a probe for researching the planets. They are also becoming increasingly significant as extraterrestrial platforms from which radiation measurements, free from weather, air pollution and noise radiation, are possible in otherwise inaccessible wavelength ranges. Astronomers are, thus, among the first utilizers of space travel for research. Modern results are not, however, only a concern of science, but also of public opinion owing to their underlying technological and masterly achievements. Examples of this are, from the USA, the X-ray satellite, "UHURU" (1970), the OAO 3 satellite, "Copernicus" (1972), with its 80-cm telescope on board and "Skylab" (1973), whose crews, apart from making scientific and medical experiments, also made observations of the Sun. Soviet astronomical satellites came under the "Cosmos" series. As examples a "Salut Station" (1974) used for research into the Sun and the Orion 2 telescope can also be mentioned here. The latter was a combination of four optical systems, three of which were used to acquire spectra in the wavelength range between 90 and 500 nm and the fourth provided X-ray pictures of the Sun. This successful venture started with a manned "Soyuz 13" spaceship (1973).

Astronomers who aim at investigating objects with less and less radiation output under conditions of constantly increasing spatial and spectral resolution very soon overtook even these possibilities and highly developed technology today really can realize projects which not so many years ago had been banned into the distant realm of utopia. Thus, in the programme of the US space authority, NASA, it is planned to put a space telescope with a mirror diameter of 2.4 m into service by the mid-eighties. As regards the conception of the optical part there is little difference from large telescopes on Earth. It is a Ritchey-Chrétien system, the radiation path of which can be directed selectively onto varied measuring fixtures. In contrast to telescopes on Earth there is no mounting at all and the whole instrument has been reduced to the tube which, in an extension behind the main reflector, houses the measuring instruments together with the maintenance equipment, control equipment and the transmitters and receivers for communication with the Earth. The large, wing-like mountings holding the solar cells which supply electrical energy are also fitted here.

An electronic camera with a similar technique as a television camera has high priority among the measuring equipment and reduces a field of sky of a few minutes of arc lateral length into 4 million picture elements and sends digital information concerning intensity distribution in its field of view to Earth. During a thirty-minute exposure, this time corresponds to the length of one "night" in an orbit at 500 km, objects can, thus, be investigated which even the largest telescopes on Earth could not photograph. For all that, even with this exposure time one can expect to reach objects of the 26th magnitude and this limit can be overcome by a superimposition of several pictures. Quite apart from range, pictures of extended sources will distinguish themselves by their accuracy in detail as a result of the lack of atmospheric turbulence. For this reason a special planet camera will also be part of the equipment and is to increase, above all, our knowledge of the distant planets of Uranus and Neptune. The plans, of course, also envisage fitting the space telescope with spectrographs and photometers. Prominence is given to an infrared photometer which by means of various radiation detectors can cover the wavelength range between 1 μm and 1 mm. This piece of equipment is cooled with liquid helium which can be kept in the conditions within the space telescope in a Dewar flask for one year.

The space telescope is planned as a long-term, extra-terrestrial observatory. By the time it is put into service the reusable space ferries, "space shuttle", will be available and will permit servicing in orbit or a return to Earth. Indeed it is planned to send a team, two or three years after launching, in order to carry out necessary checks and changes. After approximately another three years the whole system is to be brought back to Earth and thoroughly overhauled. This can be repeated several times until after about 15 years it must be finally taken out of service. As a result of such direct contact the 2.4-metre space telescope can be adapted to scientific requirements and technological progress and is, therefore, a genuine astronomical observatory.

(The annual reports regularly published by many institutions are not given in this list)

General literature:

Dorschner, J., C. Friedemann, S. Marx, W. Pfau: *Astronomy—A Popular History.* Leipzig 1975

Kirby-Smith, H. T.: *U. S. Observatories—A Directory and Travel Guide.* New York 1976

Müller, P.: *Sternwarten—Architektur und Geschichte der astronomischen Observatorien.* Bern 1975

Individual publications:

Adams, W. S.: *The Founding of the Mount Wilson Observatory.* Publ. Astron. Soc. Pacific *66,* 267 (1954)

Ashbrook, J.: *The Four Lives of a 60-Inch Reflector.* Sky a. Tel. *55,* 20 (1978)

Aslanov, I. A., G. F. Sultanov: *The Instruments of the Astrophysical Observatory Shemakha* (in Russian). Circular of Shemakha Obs., Vol. 2 (1970)

Aufgebauer, P.: *Die Astronomenfamilie Kirch.* Die Sterne *47,* 241 (1971)

Balázs, B. A.: *Das ungarische 1-m-RCC-Spiegelteleskop-Projekt.* Jenaer Rundschau *21,* 145 (1976)

Barbieri, C., L. Rosino: *The 72-Inch "Copernicus Telescope".* Sky a. Tel. *47,* 298 (1974)

Becker, F.: *Das Neue Eifel-Observatorium der Bonner Sternwarte.* Mitteil. Astron. Ges. f. 1954. Hamburg 1955

Berendzen, R.: *Zum exponentiellen Wachstum der Wissenschaft.* Die Sterne *49,* 162 (1973)

Bergh, S. v. d.: *Visiting Germany's Largest Telescope.* Sky a. Tel. *27,* 268 (1964)

Bok, B. J.: *Harlow Shapley—Cosmographer and Humanitarian.* Sky a. Tel. *44,* 354 (1972)

Bondarenko, L. N., D. Y. Martynov: *The Sternberg State Astronomical Institute.* Brief history and description (in Russian). Moscow 1973

Borchkhadze, T. M., G. N. Salukvadze: *The Abastumani Astrophysical Observatory on Mount Kanobili.* Tbilissi 1975

Bowen, I. S.: *The 200-Inch Hale Telescope. In:* Telescopes. Chicago 1960

Brouwer, D., J. J. Nassau: *The Dedication of the New Pulkovo Observatory.* Sky a. Tel. *14,* 4 (1954)

Brück, H. A.: *The Royal Observatory Edinburgh (1822–1972).* Edinburgh, no year stated

Burnham, S. W.: *Astronomy in Russia.* Astronomy and Astrophysics *12,* 595 (1893)

Cameron, R. H.: *NASA's 91-cm Airborne Telescope.* Sky a. Tel. *52,* 327 (1976)

Carleton, N. P., T. E. Hoffman: *The MMT Observatory on Mount Hopkins.* Sky a. Tel. *52,* 14 (1976)

Chappel, J. F., W. W. Baustian: *120-Inch Album.* Sky a. Tel. *14,* 178, 228, 268 (1955)

Crawford, D. L.: *The Kitt Peak 150-Inch Telescope.* Sky a. Tel. *29,* 268 (1965)

Cruikshank, D. P.: *20th-Century Astronomer.* Sky a. Tel. *47,* 159 (1976)

Cruikshank, D. P., G. Plasch: *Mauna Kea Observatory.* Univ. of Hawaii, no year stated

Dadaev, A. N.: *The Pulkovo Observatory.* Moscow 1958

Dufay, J., C. Fehrenbach: *France's Haute Provence Observatory.* Sky a. Tel. *22,* 5 (1961)

Dunn, R. B.: *Sacramento Peak's New Solar Telescope.* Sky a. Tel. *38,* 368 (1969)

Eggen, O. J.: *The Australian Commonwealth Observatory.* Sky a. Tel. *15,* 340 (1956)

Elsässer, H.: *The Project of the "Max-Planck-Institut für Astronomie" in ESO/CERN Conf. On Large Telescope Design.* Geneva 1971

Elsässer, H.: *On Astronomical Results from New Instruments and Techniques.* Observatory *96,* 224 (1976)

Evans, D. S.: *Armenian 2.6-Meter Telescope Dedicated.* Sky a. Tel. *53,* 13 (1977)

Federer, C. A.: *New National Observatory Dedicated at Kitt Peak.* Sky a. Tel. *19,* 392 (1960)

Findlay, J. W.: *The 300-Foot Radio Telescope at Green Bank.* Sky a. Tel. *25,* 68 (1963)

Findlay, J. W.: *The National Radio Astronomy Observatory.* Sky a. Tel. *48,* 352 (1974)

Forst, A. D., H. P. Palmer: *The Long-Base-Line Interferometer at Jodrell Bank.* Sky a. Tel. *32,* 21 (1966)

Fricke, W.: *Die Neueinrichtung des Astronomischen Recheninstitutes Heidelberg.* Mitteil. ARI, Ser. A, Nr. 16 (1962)

Gascoigne, S. C. B.: *The Anglo-Australian Telescope*. Endeavour *34*, 131 (1975)

Gondolatsch, F.: *August Kopff*. Mitteil. ARI, Ser. A, Nr. 17 (1962)

Gorgolewski, S., S. Grudzinska: Priv. comm. 1977

Hachenberg, O.: *The New Bonn 100-Meter Radio Telescope*. Sky a. Tel. *40*, 338 (1970)

Hachenberg, O., B. H. Grahl, R. Wielebinski: *The 100-Meter Radio Telescope at Effelsberg*. Proc. IEEE *61*, 1288 (1973)

Heckmann, O.: *Sterne, Kosmos, Weltmodelle*. Munich 1977

Heeschen, D. S.: *The Very Large Array*. Sky a. Tel. *49*, 344 (1975)

Herrmann, D. B.: *Zur Statistik von Sternwartengründungen im 19. Jahrhundert*. Die Sterne *49*, 48 (1973)

Herrmann, D. B.: *Sternwartengründungen, Wissensproduktion und ökonomischer Fortschritt*. Die Sterne *51*, 228 (1975)

Hoffmeister, W.: *Die Anfänge der Sternwarte Sonneberg*. Sonneberg 1969

Høg, E., J. v. d. Heide: *Perth 70—A Catalogue of Positions of 24 900 Stars*. Abhandl. Hamburger Sternw. Vol. IX (1976)

Howse, D.: *Restoration at Greenwich Observatory*. Sky a. Tel. *40*, 5 (1970)

Ingrao, H. C.: *News of the Soviet Six-Meter Reflecting Telescope*. Sky a. Tel. *35*, 278 (1968)

Iwaniszewska, C.: *Astronomy in Torun—Nicolas Copernicus's Native Town*. Torun 1972

Jaschek, C.: *The Strasbourg Stellar Data Centre*. Vistas in Astronomy, Vol. 21, p. 311 (1977)

Ioannisiani, B. K.: *The Soviet 6-m Altazimuth Reflector*. Sky a. Tel. *54*, 356 (1977)

Joeveer, M.: *W. Struve and Astronomy* (in Russian). Publ. Obs. Tartu, Vol. 37 (1969)

Jones, B. Z.: *Lighthouse of the Skies*. Washington 1965

Jones, D. H. P.: *Royal Greenwich Observatory—Illustrated*. Royal Greenwich Obs., no year stated

Kerr, F. J.: *Australia's 210-Foot Radio Telescope Project*. Sky a. Tel. *18*, 666 (1959)

Kerr, F. J.: *210-Foot Radio Telescope's First Results*. Sky a. Tel. *24*, 254 (1962)

Khachikian, E. Y., D. W. Weedman: *The Byurakan Observatory in Soviet Armenia*. Sky a. Tel. *41*, 217 (1971)

Kiepenheuer, K. O.: *The Domeless Solar Refractor of Capri Observatory*. Sky a. Tel. *31*, 256 (1966)

Kippenhahn, R., R. Lüst: *Max-Planck-Institut für Physik und Astrophysik, Institut für Astrophysik*. Max-Planck-Ges. Ber. u. Mitteil. 1977

Kuiper, G. P.: *Lunar and Planetary Observatory*. Sky a. Tel. *27*, 4, 88 (1964)

Kuiper, G. P.: *The Lunar and Planetary Observatory and its Telescopes*. Comm. of the Lunar and Planetary Obs. No. 172 (1972)

Kuiper, G. P.: *Lunar and Planetary Studies of Jupiter*. Sky a. Tel. *43*, 4, 75 (1972)

Leinert, V., H. Link, E. Pitz: *Die Sonnensonde Helios und ihre Experimente*. Sterne und Weltr. *13*, 86 (1974)

Lutsky, V.: *The 236-Inch Soviet Reflector*. Sky a. Tel. *39*, 99 (1970)

Mattig, W.: *Das Sonnenobservatorium auf dem Schauinsland*. Sterne und Weltr. *13*, 259 (1974)

McMath, R. R., A. K. Pierce: *The Large Solar Telescope at Kitt Peak*. Sky a. Tel. *20*, 64, 132 (1960)

Mehltretter, J. P.: *Der Flug von Spektro-Stratoskop*. Sterne u. Weltr. *15*, 44 (1976)

Meinel, A. B.: *The National Observatory at Kitt Peak*. Sky a. Tel. *17*, 493 (1958)

Milon, D.: *Helicopter Views of Green Bank*. Sky a. Tel. *48*, 360 (1974)

Minnet, H. C.: *The Australian 210-Foot Radio Telescope*. Sky a. Tel. *24*, 184 (1962)

Moffet, A. T.: *Argelander and the BD*. Sky a. Tel. *29*, 276 (1965)

Moore, P.: *A Note on the Large Russian Reflector*. Sky a. Tel. *21*, 82 (1961)

Moriyama, F.: *Tokyo Observatory Today*. Sky a. Tel. *50*, 276 (1975)

Morrison, D., J. T. Jefferies: *Hawaii's Mauna Kea Observatory*. Sky a. Tel. *44*, 361 (1972)

Moscherin, W. M.: *New Solar Telescopes* (in Russian). Erde u. Weltall (1974), p. 11

Moscherin, W. M.: *The Crimean Astrophysical Observatory* (in Russian). Erde u. Weltall (1975), p. 46

Payne-Gaposhkin, C.: *Otto Struve as an Astrophysicist*. Sky a. Tel. *25*, 308 (1963)

Philip, A. G. D.: *A Visit to the Soviet Union's 6-m Reflector*. Sky a. Tel. *47*, 290 (1974)

Plasch, G.: *Haleakala Observatories*. Univ. of Hawaii, no year stated

Preuss, E.: *Hochauflösende Radiointerferometrie (VLBI)*. Sterne u. Weltr. *16*, 86 (1977)

Ruben, G.: *Fünfzig Jahre Sternwarte Sonneberg*. Die Sterne *53*, 38 (1977)

Severny, A. B.: *Crimean Astrophysical Observatory*. Sky a. Tel. *14*, 500 (1955)

Shaw, E. N.: *Information on the European Southern Observatory*. Europ. South. Obs. 1976

Small, M. M.: *The New 140-Foot Radio Telescope*. Sky a. Tel. *30*, 267 (1965)

Smyth, M. J., H. Seddon: *The 150th Anniversary of the Royal Observatory Edinburgh and Infrared Astronomy in the United Kingdom*. Edinburgh 1973

Struve, F. G. W.: *Nachricht von der Ankunft und Aufstellung des Refraktors von Fraunhofer auf der Sternwarte der Kaiserlichen Universität zu Dorpat*. Astron. Nachr. *4*, 38 (1825)

Sultanov, G. F.: *Astronomy in the Azerbaidzhan SSR* (in Russian). Erde u. Weltall (1972), p. 49

Sultanov, G. F., I. A. Aslanov: *Brief History of the Astrophysical Observatory Shemakha* (in Russian). Circular of Shemakha Obs., Vol. 1 (1970)

Vyssotsky, A. N.: *Reminiscences of Pulkovo Observatory*. Sky a. Tel. *24*, 12 (1962)

Wampler, E. J., D. C. Morton: *The Anglo-Australian Telescope*. Vistas in Astronomy, Vol. 21, p. 191 (1977)

Wattenberg, D.: *Die Zeiss-Jena-Instrumente der Archenhold-Sternwarte in Berlin-Treptow*. Jenaer Rundschau *16*, 347 (1971)

Wattenberg, D.: *75 Jahre Archenhold-Sternwarte 1896–1971*. Archenh. Sternw. Vortr. u. Schriften No. 41 (1971)

Whitford, A. E.: *New Santa Cruz Headquarters for Lick Observatory*. Sky a. Tel. *32*, 328 (1966)

Williams, N.: *A Visit to Siding Spring*. Electronics Australia (1977)

Zhelnin, G.: *The Astronomical Observatory of the University of Tartu (Dorpat, Yuryev) (1805–1948). Historical Review* (in Russian). Publ. Obs. Tartu, Vol. 37 (1969)

Zirin, H.: *The Big Bear Solar Observatory*. Sky a. Tel. *39*, 215 (1970)

The American Ephemeris and Nautical Almanac. Washington

The Astronomical Ephemeris. London

Astronomical Yearbook of the USSR (in Russian). Leningrad

Calar Alto 1977—im Bild. Sterne u. Weltr. *16*, 407 (1977)

The Dominion Astrophysical Observatory Victoria, B. C. Victoria, no year stated

England's Great Radio Telescope. Sky a. Tel. *16*, 516 (1957)

Frontiers in Space. Washington 1967

The Hamburg Observatory. Hamburg 1964

Solar Tower Planned for Kitt Peak. Sky a. Tel. *18*, 183 (1959)

Kitt Peak's 80-Inch Stellar Telescope. Sky a. Tel. *23*, 5 (1962)

Photo Album of Kitt Peak's 158-Inch Telescope Building. Sky a. Tel. *38*, 284 (1969)

Some Kitt Peak Vistas. Sky a. Tel. *49*, 276 (1974)

The Lick 120-Inch Reflector. Sky a. Tel. *14*, 175 (1955)

Some Views of the 120-Inch Telescope. Sky a. Tel. *16*, 62 (1956)

Lick's 120-Inch Mirror Gets Final Touches. Sky a. Tel. *17*, 69 (1957)

Lick Observatory. Univ. of California, no year stated

Mount Stromlo and Siding Spring Observatory

National Radio Astronomy Observatory. Green Bank 1976

Soviet 6-Meter Telescope. Sky a. Tel. *53*, 111 (1977)

Abastumani Astrophysical
Observatory (USSR) 10, 11,
12
ADN Zentralbild, Berlin 9, 17,
18, 87, 121
Aerofilms Ltd.,
Borehamwood 58
APN, Moscow 1
Archenhold Observatory,
Berlin-Treptow 14, 15, 16
Artus, H., Jena 8, 74, 96, 97
Asiago Astrophysical Obser-
vatory (Italy) 13
Astronomical Calculating
Institute, Heidelberg 51
Astronomical Institutes of
Bonn University 19, 20
California Institute of Tech-
nology and Carnegie of
Washington 64
Central Institute for Astro-
physics, Sonneberg Obser-
vatory (GDR) 92
Crimean Astrophysical Obser-
vatory, Simeiz 88, 89, 90
Dominion Astrophysical
Observatory, Victoria, B. C.
118, 119, 120
European Southern Obser-
vatory, Munich 70, 71, 72
Geiges, L., Freiburg (Federal
Republic of Germany) 44
Hale Observatories, Pasadena,
California 73, 75, 76, 77
Hamburg Observatory 49, 50
Harvard College Observatory,
Cambridge, Mass. 25, 26,
27, 28
Häusele, I., Sonneberg (GDR)
91, 93, 94
Haute-Provence Observatory,
Saint-Michel (France) 80,
81, 82
Herrmann, Dr D., Berlin 57
Institute for Astronomy,
Honolulu 62

Institute of Astrophysics and
Atmospheric Physics of the
Estonian Academy of
Sciences, Tartu (USSR) 79,
98, 99, 100
Karl Schwarzschild Obser-
vatory, Tautenburg (GDR)
101, 102, 103
Kiepenheuer Institute for
Solar Physics, Freiburg
(Federal Republic of
Germany) 45, 46
Kitt Peak National Obser-
vatory, Tucson, Arizona
115, 116, 117
Konkoly Observatory,
Budapest 23
Kroitzsch, Dr V., Potsdam 24
Lick Observatory,
Mt. Hamilton, California
83, 84, 85, 86
Linde, G., Jena 4
Lunar and Planetary Labora-
tory, Tucson, Arizona 111,
112, 113, 114
Main Astronomical Obser-
vatory, Pulkovo (USSR) 78
Max Planck Institute for
Astronomy, Heidelberg 52,
53, 54, 55, 56
Max Planck Institute for Radio
Astronomy, Bonn 21, 22
McDonald Observatory,
Fort Davis, Texas 40, 41,
42, 43
Mount Stromlo and Siding
Spring Observatories,
Canberra 32, 33, 34, 35,
36, 37
Muzeum Okregowe, Torun
(Poland) 108, 109
NASA Ames Research Center,
Moffett Field, California 65,
66
National Radio Astronomy
Observatory, Green Bank,
West Virginia 47, 48

Nuffield Radio Astronomy
Laboratories, Jodrell Bank 63
Observatoire Astronomique,
Strasbourg 95
Observatory of Jena Uni-
versity 24
Observatory of Nicolaus
Copernicus University,
Torun (Poland) 110
Royal Greenwich Observatory,
Herstmonceux 59, 60, 61
Royal Observatory, Edinburgh
38, 39 a and b
Smithsonian Astrophysical
Observatory, Cambridge,
Mass. 29, 30, 31
Special Astrophysical Obser-
vatory, Zelenchukskaya
(USSR) 122
Sternberg State Astronomical
Institute, Moscow 67, 68, 69
Tokyo Astronomical Institute
104, 105, 106, 107
Vogel, Prof. Dr W., Jena 5